WITH COMPLIMENTS FROM THE AUTHOR

Carl Zimmerling

ASTRONOMY FOR THE LAYMAN:

REALITY BEFORE BELIEF

Carl Zimmerling

Spiritual Fulfillment
through Common Sense
and Basic Science

RAINBOW BOOKS, INC.

Library of Congress Cataloging-in-Publication Information

Zimmerling, Carl, 1931 -
Astronomy for the layman : reality before belief / Carl Zimmerling
p. cm.
Includes index.
ISBN 1-56825-026-6 (alk. paper)
1. Astronomy. 2. Religion and science. 3. Bible and science.
I. Title
QB43.2.Z56 1995 95-21296
215--dc20 CIP

Astronomy for the Layman:
Reality Before Belief

Published by Rainbow Books, Inc.
P. O. Box 430
Highland City, FL 33846-0430 USA
Editorial Offices Telephone (813) 648-4420
Orders: Telephone (800) 356-9315, Fax (800) 242-0036

All photos and illustrations provided by the author.

Interior Design: Marilyn Ratzlaff
Cover Design: Betsy Lampé

Manufactured in the United States of America.

CONTENTS

SECTION II — 249

SECTION III — 339

INDEX — 363

PREFACE

Regardless of age, race, religious background or level of education, millions of inhabitants of our planet have been educated by two usually opposing systems. Young people in particular are easily confused by the vast differences between "religious" education and "worldly" schooling. Starting with the youngest children, all are effectively and often purposely misinformed about scientific reality through early religious education [bible schools], as well as through "inheriting" a religion from their parents and forefathers. Every particular culture and the religious tradition of their ancestors are passed on from generation to generation, usually without questioning the logic and consequences of related actions.

As a result of unquestioned acceptance of religion, of tradition and because of general lack of knowledge about basic science, (also in order to comply with the belief and wisdom of their elders), many youngsters and teens are torn between "belief" in religious teachings and the "facts" of modern science. For them it is, as it was for the author at times, impossible to be a "good" Christian, a "good" Moslem, a "good" Jew or a "good" Hindu, etc., and at the same time accept all scientific FACTS and LOGIC, twentieth century discoveries and fact-based theories . . .

All the facts of nature are before us. Only many, maybe most have not yet been discovered. Sciences are still very, very young and our present knowledge is a mere drop from the well of total potential scientific wisdom. Mankind can really only count on the past 100 years as far as major scientific progress is concerned. And, what little we know by now is often either overlooked or purposely "swept under the carpet". Many major discoveries are often belittled or completely ignored by religious leaders all around the world, whenever such discoveries squarely contradict religious teachings.

After a brief introduction to "BELIEF" and "FACTS," I will address all readers, young and old, who are undecided between ancient religious belief and twentieth century knowledge. To all of those readers, who have a lot of questions on their mind, I will introduce many little-known facts about past and present religious extremism, biblical prophecies and the sometimes bloodstained history of major world religion's teachings. The legal system when it is based on religious tradition will be examined for logic.

Throughout the book, I will provide researched examples of well-known as well as little-known scientific discoveries and fact-based theories. The material and scientific reports used for research were usually readily available to everybody from daily newspapers and magazines. But new discoveries are frequently overlooked, as they often appear in a hidden corner of the last pages. It seems to be particularly noticeable, when the information contradicts religious teachings and it does not make good "News."

I will compare a number of scientific findings and logical fact-based theories to religious scripts and prophecies by applying simple common sense. The human body and brain will be compared with the rest of the animal world. Emphasis will be given to "when and why" best to decide between FACT and logical theories, vs. ancient and blind fundamental BELIEF. I will provide believer and non-believer with valuable information and a lot to think about.

One chapter will examine the human "soul" (or "spirit") and the necessity of religion will be questioned altogether, whenever scientific facts clearly contradict ancient holy books and scrolls.

After examining the results of various religious (or holy) wars, the question will come up repeatedly: Why apply conflicting ancient religious teachings and prophecies, when humanity has the realities of the twentieth century at hand? The results are often devastating to believers and to non-believers equally.

ASTRONOMY and related sciences to the author are the only realistic sources of information about the TRUE origin of life and the Universe, and about the REALITIES of evolution, which has an undeniable link to all the laws of nature. A major part of "Reality . . ." is devoted to the facts and theories of Astronomy.

I will try to introduce scientific realities and theories to the readers in a most "non-scientific" manner, and as an "amateur scientist," to as many other "amateur scientists" as I can possibly hope for. All readers who have some doubts about the purpose of major religions, and with basic interest in sciences, should benefit from an overall view of everything of importance to humanity. Or at least, what SHOULD BE important and what should be general knowledge.

The author will ask in one form or another, and more than once, "what has religion done for mankind, past and present and in contrast, what is scientific knowledge doing for us?" The results are obvious. Over the recent centuries, mankind has seen the consequences of the acceptance as "fact" of religious "legends" and "belief". At times, some actions were understandable because very few "real" facts were known. The author hopes, with the help of twentieth century sciences that this book will demonstrate the all-important differences between "Belief," "Facts" and "Logic."

In one chapter, as well as throughout the book, examples will be introduced about the most creditable eyewitness reports of sightings and encounters with so-called UFOs and their obviously technologically vastly advanced extraterrestrial pilots. Only the most creditable, usually multiple encounters have been chosen, those which do not deserve to be written off as hoaxes. Examples have also been selected from some of the most remarkable reports of superior knowledge of astronomical facts by some ancient civilizations. Some of them are dating back to periods long BEFORE our major religions misinformed humanity about creation and the universe, mostly during the past 3,000 years.

In the final chapters I will concentrate on the priorities of today, of what modern science and knowledge can do to change our planet into a better place to live, (at least in my opinion) and with as little religious influence as possible. The reader will have a glance at our world as to how it COULD appear to hypothetical extraterrestrial visitors in the first century of the new millennium. And IF humanity has chosen the priorities for the immediate future correctly, by applying a maximum of common sense.

Throughout the following chapters reference will be made frequently to Thomas Paine, the famous author of *Common Sense* and *Age of Reason*. For those who are not familiar with this historical name, Thomas Paine was most instrumental in the writing of the Declaration of Independence and the height of his career was about 200 years ago. The reader will find his quotes throughout this book because of his at the time far advanced opinions about religion, common sense, the freedom of speech and from religion, for that matter. After near-completion of this manuscript, it appeared to me that he often used almost the identical words and opinions two hundred years ago as I did at the end of the twentieth century!

SECTION I

CHOOSING PRIORITIES: Ancient Religious Myths or Twentieth Century Science?

It happens to all of us one time or another and we give it little thought. Before we reach for our wallet, we guess the approximate amount of money which we BELIEVE the wallet to contain. Let's say this time, our guess was "about" forty dollars. We could leave it at that and say, "I BELIEVE, I have forty dollars on me."

Up to this point, we are dealing with BELIEF. If we want to have FACT, we would now empty our wallet and count the money. Let us say, we were fairly accurate and it ended up to be $41.36. Now we have FACT. Another example would be one's balance in the bank account, a wild guess would be "I believe I have $1500 in my account." The search for "Fact" would be rather shocking when the actual balance is an overdraft of $457.09. It happens to the best of us sooner or later, one forgets about a postdated check, for example. Up until we discovered the "reality" of the overdraft, "belief" put our mind at ease, until we took our "head out of the sand" and discovered the "facts."

One could take this little experiment one step further and take a wild guess on how many stamps are in a friend's stamp collection. First of all it would have to be a very wild guess. In this case, the collection might contain maybe 2000 to 3000 stamps, looking at a box full of loose stamps and a few albums.

Since it is a friend's stamp collection, who is hardly ever at home, we have to leave it at that and BELIEVE that he has 2000 to 3000 stamps in his collection. When we talk to others about his stamp collection, we would refer to it in hypothetical, "believed" numbers. In this case it is much easier to BELIEVE than to search for FACTS. Maybe some day we will actually get together and count the stamps, and only then will all concerned have FACTS with actual numbers. In the end, the actual count could be way out at 4500 stamps, or we may be close with 3100 stamps.

In real life, and quite often due to lack of scientific knowledge, humanity had to BELIEVE in what holy books such as the Bible and the Koran taught hundreds or thousands of years ago. There was no opportunity to verify or to discredit religious teachings until recently. Our ancient ancestors had no choice and they had little reason to doubt the scriptures, or they were conditioned by their all-powerful religious leaders to be afraid to do so.

In the following sections and chapters, the reader will have an opportunity to examine and to understand the FACTS and laws of nature, the realities and fact-based theories of astronomy. We will look at humanity, "homo sapiens" as basically just another mammal, in some ways superior, but in many ways very inferior to other creatures. We will examine closely what we have in the past BELIEVED about ourselves, our mind and body, how we are created, and how and why we and other creatures die.

The reader will be introduced to little-known discoveries, often verifying the remarkable knowledge about the universe by ancient civilizations and by rather obscure native tribes, long before modern science could have discovered the facts. The author will compare the probabilities of "reality" and "logic" from ancient holy teachings versus modern day eyewitness reports of encounters with "alien" spacecraft, (UFOs).

There will be a critical view of what millions still BELIEVE, even though contradictory scientific facts are available to all. Facts on hand usually contradict religious myth, more often today than ever before. It will become clear to the reader how often religious believers tend to ignore obvious realities of nature's laws, and WHY they choose to do so. It will be apparent how often LITTLE attention is paid to proven discoveries by astronomers, and how MUCH attention is paid to unproven ancient religious teachings at the same time.

Although religious teachings and scientific facts DO occasionally agree when they are based on pure logic and on common sense, it is rarely the case, there will be a number of examples. It may soon appear obvious to the reader how much more questionable ancient religions have become, compared to the twentieth century's scientific knowledge. Yet, there is still a lot to be discovered. It will become clear also how often the DATE of writing of a scripture is the only logical explanation for religious doctrine, and WHO the writers were.

One chapter will deal with Homo Sapiens, the "mammal," human and animal intelligence, instincts and emotions. Human feelings of love, lust and our reproductive system will be compared to other animal reproductive systems and all creatures' natural sex drive. The reader will be surprised how similar many human body and brain functions are as compared to those of other animals.

A look at traditions and religions in general and at specific religiously inspired criminal laws and traditions will follow. The benefits derived from religious belief will be compared with regression, death and destruction created by it—all in the name of often blind religious belief.

On the bright side, the possibilities of "co-existence" between traditional religious belief and scientific facts, logic and common sense will be investigated. On the dark side, it will also appear obvious how discoveries through medical science are ignored, in favor of traditional religion.

In Section II, the reader will be introduced to the facts and theories of astronomy in the least possible "scientific" manner. Astronomy is an often misunderstood science, sometimes even confused with astrology, and frequently ignored and overlooked because of the highly scientific way in which it is usually introduced by professional astronomers and physicists. As mentioned, text books on astronomy are usually written for other scientists and for science teachers.

It may become more obvious to the reader how many difficulties astronomers, physicists and other scientists still have today when introducing scientific reality. In order to even introduce a controversial but fact-based theory, scientists have considerably more "selling" to do than theologians will ever have with biblical prophecies and quotes.

Anything that cannot be explained by logic in religion is written off to faith and to unquestioned, traditional belief. Even if a discovery is proven to be scientific fact, it is usually easier for deeply religious believers to simply stick to an old myth. In the past, it sometimes took hundreds of years for our ancestors to introduce and accept technological advances and scientific discoveries, as they were contradicted by deep-rooted religious superstition, particularly when such discoveries disagreed with old holy books. And it is still happening today.

Only during the past fifty or so years, our astrophysicists and astronomers have discovered that the entire universe is basically made up of the same matter as the solar system. The same atomic material, which can be found within the Earth and our Sun, is also present everywhere else in the universe. The same laws of physics and chemistry apply everywhere. Thermonuclear (fusion) reaction within our Sun is identical to all other visible stars of similar age.

Our ancestors were not aware of the incomprehensible number of stars in the universe and religious scripts have not taught us anything about it. Humanity found the answers without help from holy books. Astronomers are discovering more new galaxies with great regularity in a universe of such vast dimensions that even the most open-minded scientists have difficulty in comprehending it. Gone are the days when the Earth was the center of the universe. Mankind no longer has to BELIEVE in a flat Earth — we have the contradicting scientific FACTS. We will look at those and other "facts" and compare them with ancient religious belief throughout the following chapters.

The last Section will be primarily a summary. The author will make reference to human ignorance of the vast scientific knowledge which has been accumulated particularly during the past 100 years, by continuing with the "status quo." Possible priorities will be introduced on how humanity could make use of scientific knowledge in all fields, and how common sense should rule our lives, as a priority over outdated religious myths.

Without going into wild utopian fantasies, the author will look critically at some of our most common current legal systems. The necessity of increasing the application of current scientific knowledge will be emphasized, together with a look at how hypothetical, advanced extraterrestrial observers may possibly see planet Earth today. The reasons for separation between elected governments and religious "leadership," (between church and state) will be stressed. The author will introduce possibilities of how much more humanity may be able to gain from scientific research, given as little as another hundred years, now that the greatest obstacle of this century, the cold war, has been eliminated.

HOW MUCH REALITY CAN BE FOUND IN RELIGIOUS BELIEF?

It is said that God created Man in His image. Those who believe it must have asked themselves often, was the "Creator" the familiar image of an old man or was it perhaps even a woman? Was the "Lord," as He is often referred to, an Asian, an African, or perhaps a native Indian? Or is "He" a male Mediterranean, white Caucasian, with the appearance of a "God" as He is often illustrated in religious books? Who came up with the idea of the bearded white male "GOD" in the first place? Has anybody during recorded history actually "seen" a God? Or I should ask, has a large group of people seen Him in person AND did they have a good sample demonstration of His powers?

Is it not more logical to look first at the FACTS around us and observe reality and read about the Universe and the sciences, before we blindly accept all ancient religious teachings and scripts?

If God is a white Caucasian, how do we account for black Africans or for that matter, where did the world's very old Asian civilizations come from? Where did the Mayan civilization in Central America come from? Their history can be traced as far back as 5,200 years. Other African and Asian tribes and cultures can be traced back much further to times long before the single "Creator" theological theory was introduced to, and sometimes forced upon humanity. Why would "God" wait so long to introduce His existence to humanity only a few thousand years ago when "modern man" really existed long before that? Why only to the population of Palestine and the surrounding Mediterranean lands? Civilized cultures existed at that time in Asia and America in much greater numbers than in the land of the Bible.

An increasing number of nonbelievers and believers throughout the world, and of various inherited religions, no longer believe in the bearded old (white?) man as an image of a "God." More and more of them question most religious teachings in general and now, more often than ever before, many of them DO use scientific facts, logic and common sense instead. But too many simple religious "believers," cult members, fanatics, fundamentalists and statesmen still hang unto ancient, illogical tradition for guidance.

A steadily decreasing number of followers from within the population of technologically advanced nations still regularly attend houses of worship. A drive through urban areas in Europe, Asia or in North America during the respective weekly "holy day" or "day of rest" will verify

A delightful sight indeed — in an "anytown," typical North American community. Whatever the cost was for the three "houses of worship," all were built or remodelled during the recent recession. Worshippers may now be more comfortable worshipping an unknown, mythical "God." At the same time, local food banks were desperately pleading for more donations, many locals found themselves homeless. WHAT WERE THE "FAITHFUL" WORSHIPPERS' AND DONORS' PRIORITIES? A typical example of illogical "Belief over Reality."

this statement.

The decline in church attendance is no secret and has been admitted by religious leaders. Only certain religious sects, often with questionable or dubious goals, seem to be successful in attracting increasing numbers of new followers. Most other major religions have lost followers at a steady pace.

As a typical example, based on 1990 statistics, 89% of the Ontario, Canada population of about 8,500,000 reported to be members of one or the other Christian faith. Assuming 89% also applies to the inhabitants of Toronto, about 3,115,000 believe in Christian religions in that city. With a hypothetical 90% attendance, local churches would have to be able to seat about 2,800,000 faithful Christians every Sunday in that city. Of course, reality is far from that figure.

Most major religions HAD to alter their views and revise their teachings, at least trying to keep up partially with the realities of the past 100 years. Without some changes, based on simply overwhelming scientific facts, many religious communities would have lost an even greater number of followers.

Forceful leaders have emerged and founded (sometimes dangerous) religious cults, often with thousands of blind followers, blind to the point of committing suicide. We had a good example with the Jim Jones event, when about 900 followers poisoned themselves with cyanide-laced soft drinks in Guyana. Cults seem to be the only religious groups now, where new members are successfully recruited. Obviously, THEIR translation of scriptures is more attractive to the new followers. Unfortunately, the number of contradictions coming from holy scripts are so numerous, making it easy for a cult leader to pick the prophecies and verses from the Bible, which suit HIS ambitions. The followers are often uninformed about scientific facts and scientific theories, which originate from logical conclusions. They therefore fail to apply religion logically.

Most well-known scientists do not believe in the human image of a "God." As one example only, Dr. Carl Sagan, a world famous American professor of astronomy and a science writer was asked that question at a seminar at the University of Toronto. As could be expected, his answer was based on facts and logic. No, he does not believe in the biblical image of God and in the teachings about creation. His answer was to the effect, that if a "creator" really exists, it must be the greater Universe itself, with the same laws of physics, chemistry and nature set firm and applicable anywhere in the cosmos.

Long before humans, "Homo Sapiens" began to walk this planet, we had other vertebrates living on Earth. Later the mammals and primates evolved and we are modern versions of this group of creatures. All living organisms, walking, crawling, flying or swimming on this planet along with plant life, are shaped by climate, genes, mutations, atmospheric changes, and by the elimination of the weakest and survival of the fittest over hundreds of millions of years. Short of causes through a major catastrophe, evolutionary changes are never fast in nature's system. We have overwhelming evidence to that fact. A tremendous number of early natural catastrophic events wiped out all species which could not adapt fast enough. Some of them survived and we are here today as a result of it.

Primitive life evolved in simple forms on Earth over 4,000,000,000 years ago, from the prevailing chemical and physical conditions on our planet at that time. Along with the effects of solar radiation, the entire surface of our planet was drastically different from today's. And conditions have changed ever since and will continue to be changing as long as the solar system exists.

In order to comprehend these evolutionary changes in nature for ALL forms of life, the emphasis is on "hundreds of millions of years," and "billions" of years. We must try to understand astronomical time and forget about one or ten or even one thousand generations of mammals and other life forms.

As an example, let us take any primate with a life span of 70 years, such as homo sapiens himself: based on archaeological excavations, near-human, apelike primates have existed as far back as nearly five million years ago. Five million years translates into more than 71,400 lifetimes of 70 years each. The slightest change in appearance over each generation would result in an almost entirely different species during 71,000 lifetimes, or an equal of about 250,000 generations later. And this process will go on for the next 250,000 generations, failing a major environmental, disastrous change. It has to be more than coincidence that chimpanzees have 99% of human genes . . . is it not enough proof that we all originate from a common family of prehistoric ancestors?

I am referring to the "Australopithecines," of which remains have been unearthed mainly near the Olduvai Gorge in Tanzania. Their appearance was amazingly close to modern man and they may have been the first upright walking primates. A sensational discovery was made when archaeologists discovered a half skeleton in Ethiopia, the famous "Lucy." Unfortunately, and as usual with such revolutionary discoveries which could contradict biblical, traditional belief, the importance of this event was also downplayed in the media and fundamentalists tried to discredit it as a hoax.

FOUR BILLION YEARS SQUEEZED INTO 24 HOURS

One better way to comprehend the age of our planet and to understand how long it took for life to develop through evolution, is to imagine the Earth to be 24 hours old. During the first 6 hours our planet was extremely hostile and volcanic, worse than the planet Venus today. Nearing the end of the first 6 hours, an atmosphere had developed, but still of an entirely different composition, compared to today's thin layer of 78% nitrogen, 21% oxygen and 1% miscellaneous gases.

Toward the end of six hours, conditions became acceptable for very primitive life to develop, all the chemical requirements for life fell into place. An abundance of molten rocks was supplying the necessary chemical components for DNA-type bacteria to form. Single cell microbes turned the seas bluish and another single cell organism, algae, evolved as the first plant life. By the way, placing even DISTILLED water into a sterile container and by exposing it to sunshine for a few days, simple algae will actually evolve NOW, under today's atmospheric conditions — "creation" at work. The laws of nature cause algae to form the chlorophyll base for other plant life.

Given the same circumstances, there is no reason to believe, that life would not form on ANY planet in the universe, other than Earth. However, one must remember, no two stars, planets or natural satellites are EXACTLY identical, no more than two people are an EXACT copy of each other, not even identical twins. No two trees are precisely identical anywhere on Earth.

Single-cell life started to divide during our planet's age of between six and nine "hours." Again, conditions had to be right for this action to take place. Thus, trilobites and multiple cell creatures developed and the first very primitive insect forms appeared, initially in the oceans and much later on land. This process alone, for "cell colonies" to form, took another "three hours," or getting back to reality, an incredible period of about 500,000,000 years.

By now, conditions on Earth were ideal for all life forms to evolve further. After 10 hours, we had the first back-boned animals. After about 10 and one-half hours, plant life was abundant, thanks to ideal climatic conditions at that time, and it then became the foundation for more rapid development of animal life. At the same time, the continents started to break away from one "super continent," and by 11:00 the first small primitive reptiles evolved from an earlier amphibious stage with lungs.

More dinosaurs evolved as individual species forming colonies, roaming Earth for another 100,000,000 years — much longer than humanity. Yet by our 24-hour clock, the entire age of dinosaurs lasted less than an hour. At the dramatic end of that period, other smaller mammals had a better chance to survive the natural catastrophe of the day. One of the surviving smaller mammals became the ancestor of homo sapiens and that of other primates.

During the LAST FEW SECONDS of our clock, the first primitive version of man left their bones and hand tools for our modern archaeologist to find in the twentieth century. We have most of the evolutionary evidence of the entire process available to us now, but with very little help, if any, from ancient religious scrolls and books.

To summarize the development of life on our spaceship Earth's journey through the cosmos in more "real" time (in millions of years):

4,000 — single cell life forms, such as algae and bacteria evolve,
550 — mostly arthropods, ancestors of today's crustaceans, insects and spiders,
530 — Anomalocaris was the giant of its days at about 3 feet in diameter, remotely resembling the horseshoe crab,
420 — the first jawed fish appeared,
360 — vertebrates start to colonize the land,
240 — dinosaurs start to emerge,
210 — about 1000 species of dinosaurs on Earth,
60-70 — the age of mammals begins, about the same time when the dinosaurs disappeared.
4 — homo erectus, later Homo Sapiens appear.

Compared to the age of this planet, the existence of even the most primitive multi-celled life-forms started only during the last one-eighth to one-ninth of the age of the solar system.

EVOLUTIONARY ANCESTORS OR DESCENDANTS OF ADAM AND EVE?

More often now than ever before archaeological discoveries appear to bring us closer to proving the evolutionary development of land animals, (including the human mammal) from common reptile-like creatures. These distant ancestors date back hundreds of millions of years, when planet Earth was not at all what it is today. It must be emphasized, primitive "humanoids," not really humans and not really apes evolved finally "only" about five million (not billion) years ago. "Modern" man's ancestors only dates back about 100,000 years, in comparison. However, we have changed our appearance somewhat, even during this relatively short time. As an example, most humans are taller now than our ancestors were only 500 to 1,000 years ago. We are still subject to change and will be, as long as humanity can survive at all on this planet.

At this point, I want to focus on another ancestor, the "Ramapithecines," which date back even further than Australopithecines. Although only bone fragments had been found in Africa and Asia, scientists were able to establish that their teeth were relatively close to ours. We therefore can easily come to the conclusion, that their diet at least was very similar to ours. This most ancient possible ancestor of homo sapiens lived on our planet between fourteen and eight million years ago. How does that compare with a close resemblance of "modern" man and its existence of less than 100,000 years? Unfortunately, very little is known at this time about the "Ramapithecines," since primarily only skull and jaw bone fragments have been found. Nevertheless, an exciting discovery.

Another finding in Africa some time ago revealed remains of a primate, the twenty million year old "Kenyapithecus Africanus." During that prehistoric time period, the ancestors of apes lived together with our own ancestors and both were using primitive hand tools. It was established back in 1969, based on this discovery that man did not come from ape, but both man and ape have the same ancestry — somewhere in the much more distant past. Twenty million years ago, both man and ape were "plain" dwellers. While man remained on the ground, some gorillas later took a preference to life in the trees. Today, religious believers still spread the misconception that evolution teaches that man came from apes.

Once again, even the slightest change in our appearance, body functions and brain development during ONE life time would multiply astronomically over, say, 70,000 life times. Throw in the odd mutation, and we can account for the remarkable changes to the appearance of "humans" over a few million years. Let us keep in mind that other life forms existed as far back as 400,000,000 years on this planet, but under completely different climatic conditions. It is important to remember once again, even the chemical composition of the air which we are breathing now, was entirely different and all life forms developed accordingly. Considering that our biological ancestors date back at least that far, their existence would have been 5,700,000 "human life spans" ago.

The close relationship between man and ape is being confirmed with the additional discoveries of skeletal remains of ancient, more common ancestors and through study of genetic material. A few logical theories suggest that orangutans separated gradually from our common ancestors about sixteen million years ago, gorillas diverged later around about ten million years ago, and chimpanzees and "hominids" parted very likely between six and seven million ago. The first "Homo erectus" appeared three and one-half to two million years ago in Africa.

While "Peking man" and "Java man" evolved parallel on the Asian continent between one and one and a half million years ago, other ancestors of modern man evolved further in Africa at about the same time. Only our African ancestors had been around much longer, as the 1974 discovery of the remains of a (female) "Australopithecus Afarensis" has proven, better known as the partial skeleton of the three-million-year-old "Lucy."

The tiny female weighed a mere 65 pounds and was

only three feet, eight inches tall. According to Bone structure she had walked mostly "erect" on the ground, contrary to apes. The more closely related apes of her time preferred life in the trees, (probably less dangerous) and walking on all four limbs.

A typical result of evolution was the Neanderthal Man, remains of which were found in Europe. He roamed this part of the world unchallenged by other human "cultures" between 200,000 and about 35,000 years, much longer than our modern civilizations exist. His body had adapted to the cold and harsh conditions of the on-again off-again ice ages in Europe by being short and stocky, (better to retain body heat), with considerably more body hair than any other "hominid" ancestor. For reasons unknown, the Neanderthal man disappeared 30 to 35,000 years ago, after the more intelligent and nomadic modern man met with him. Modern "homo sapiens" originated from the more moderate "Mediterranean" climate and may have exercised his "warrior skills" for the first time, by exterminating the more primitive Neanderthal tribes?

Very recently, archeologists discovered a shin bone of what they estimated to come from another erect walking human ancestor. The discovery was made in Sussex, England about 25 ft. underground. Considering the location and the estimated age of the bones of 500,000 years, this six foot (plus) tall early human may have been the ancestor of the more recent Neanderthal Man. With the on-again off-again ice ages covering Europe at the time, the six-footer shrunk in size as protection from the cold climate and the "Sussex" man may not have had as much body hair as the Neanderthal Man for that reason.

To sum up this brief history of human origin, the most remarkable conclusion is that our ancestors evolved EVERYWHERE on this planet, in Africa, and in America (40,000 yr. old skeletal remains have been found here), in Australia (50,000 year old remains) and in Asia. Biblical history seems to ignore these facts completely by concentrating on the Middle East only. It is most noticeable to the critical observer.

Getting back to the millions of years before "Lucy," remains of advanced, non-humanoid life-forms, such as more "advanced" dinosaurs, have been recently unearthed and are estimated to be about 220 million years old, give or take a few million years. Again, we must try to comprehend a period of "220,000,000 years," it translates, to about 3,143,000 human lifetimes of 70 years. Maybe that figure is easier to understand, comparing it to our modern calendar which dates back 2000 years; multiply this time period of most of human history by 110,000, and we have 220,000,000 years!

ANOTHER THEORY ABOUT THE ORIGIN OF MODERN MAN

If our ancestors had originated from another solar system and were stranded on Earth, say four million years ago, the natural evolutionary changes to their bodies and brains would still have been most significant, through environmental conditions and through occasional mutations. If we had a parallel development of earthbound Homo Sapiens, a mixed race of part alien and part "earthling" could have resulted. It is highly probable, but to this date without scientific evidence. Many scientists believe in this theory in order to explain the so-called "missing link" in brain power, to close the apparent intelligence gap between pre-historic humanoids and the highly advanced civilizations of the past ten thousand years. Also, natural disasters had and still have, a major effect on all local life-forms. That is, if such life is not completely wiped out by these events.

Whether we like it or not, and whether we originate from reptile ancestors on this planet or from another solar system, or from a combination of both, we are only ONE species of the family of vertebrate, and more precisely, one of many thousands of species of mammals.

More specifically, we are primates. Regardless of religion, race or tradition, human races on Earth have the same BASIC appearance, the same body forms and functions, the same digestive system and the same means of voice communication. Humans everywhere have the same capability of using their hands as necessity for survival. All races are effected by the same diseases and causes of death, with only a few minor exceptions.

As we can see, ancestors of homo sapiens have been around in one form or another at least five million years, maybe even fourteen million. Modern religions on the other hand appeared only 4000 to 5000 years ago. Prior to that, only tribal religions and traditions were practiced. Many tribes worshipped the sun as the almighty source of life. In a way, it is not so far from the truth. At least, it is an established fact that life without the Sun would be impossible on any planet. Papago Indian tribes also have recognized the importance of solar power, as many villages use solar power as the only power source, even today. It took the rest of humanity to the second half of the twentieth century to concentrate more on the power of the sun, in the light of devastating results from the use of other energy sources.

Although it is against common sense to pray or to sacrifice lives to the sun, as practiced by some ancient civilizations, traditional worshipping of the sun by many native peoples is based on reality and to thank the mighty sun in one way or another. To them, there is no doubt about it, without the sun life on Earth could not exist. That is a fact and it has been known to our ancestors long before Judeah, Christian and Islamic religions appeared on Earth. On the other side of the coin, how much PROOF do we have that a hypothetical image of man, a single "GOD" is the creator of life, and not the sun?

There is no evidence of MAJOR wars between civilizations during all the millions of years of the existence of homo sapiens on Earth, BEFORE major "modern" religions were introduced about 4000 years ago. Coincidental with the appearance of our major religions over the past 3000 years, we also have had the greatest loss of lives and the most destructive conflicts. Much of it simply because of DIFFERENCES in religious beliefs.

The world is certainly NOT a better place through religion than it appears to have been millions or even as short as 10,000 years before that. If anything, with stronger religions and more fanatical belief, the hostilities seem to have increased to this day. Common sense will prevail, only if and when humanity realizes the importance and the difference between blind belief and scientific facts and when criminal laws and politics are no longer more or less based on traditional religious faith. Our decisions will have to be based first and foremost on REALITY and FACT-BASED LOGIC, and on religious tradition and ancient scriptures ONLY, when common sense can be applied to it.

THE REALITY: ONE BIG GLOBAL ANIMAL FAMILY

All mammals and most vertebrate not only multiply in the same way, they all have the same basic body forms and anatomical parts. Alligators, cat-family animals, elephants, members of the bird family and apes, all have the basic two eyes, two lungs, two kidneys, a brain, four limbs, a rib cage, and a spine. And most mammals have a similar body temperature — their blood is of the same color and of similar chemical substance.

Embryos in the very early stages of development look so similar that it is almost impossible to tell the difference between a human embryo after about four weeks of development and that of a fish, a turtle, a bird or any mammal after an equivalent time of development. The digestive systems are basically the same, we must breath in and out the same air mixture as other mammals do, we nurse our babies the same way. Our bodies respond the

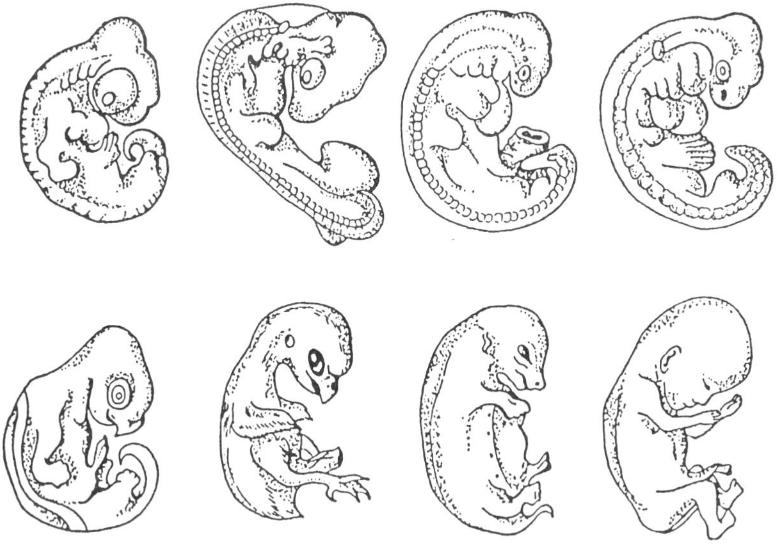

Illustrated: something to think about for those, who believe that man is "superior" and different — Embryos of rather distantly related creatures: Top left to right — early embryonic stages of turtle, chicken, pig, man. Bottom: the same embryos, slightly more developed — proof that MAN IS NOT SPECIAL, contrary to what scriptures say.

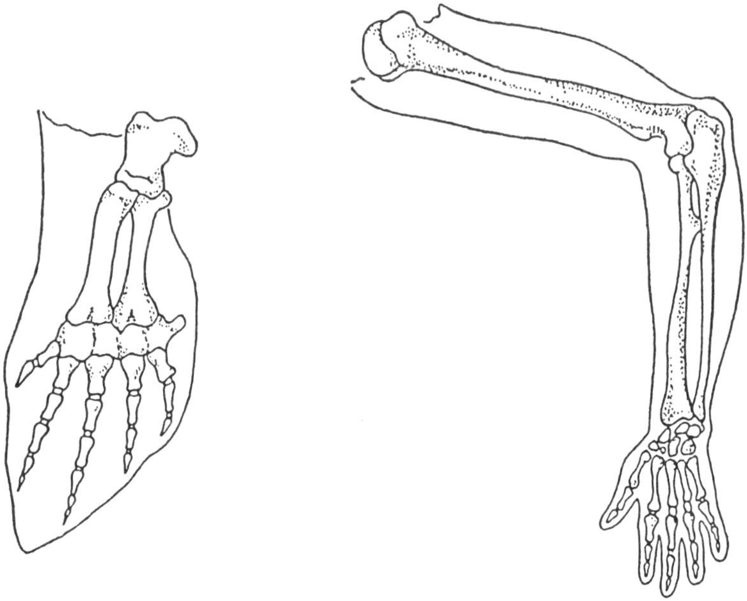

Illustrated: two distant relatives — but their "hands" and "arms" are still remarkably similar . . . whale and man.

And: the feet (left to right — chimpanzee , forest gorilla, mountain gorilla, Homo sapiens (man) — all primates ARE closely related.

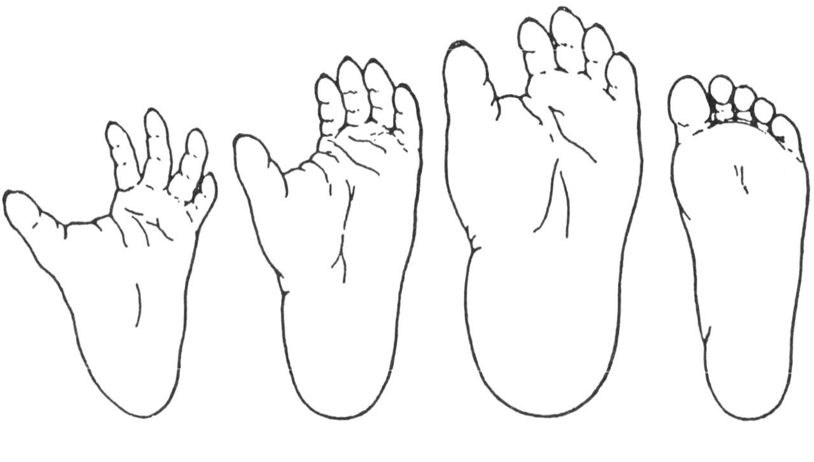

same way to environmental changes (frequently man made), sometimes resulting in mutations and malformed babies.

However, WE are much more sensitive to diseases than most of our fellow mammals, and many of them are caused simply by the wrong, unnatural nutrition, or by man-made pollution. Let us look now at humanity's "superiority" over our "fellow" creatures, (as it was written in ancient religious scripts some thousands of years ago).

What sets us apart from other mammals? Again we have to look at the FACTS. We may THINK we have superior brain power, are capable of logical reasoning and we do have highly developed vocal cords for communication. We can use our two hands with much greater sophistication than any other animal. Is it a gift of "God," or is it because the human body is the WORST developed of them all? Facts point to the latter. Religion has taught us, that only man can think; this assumption has already been doubted by the inventor of the computer himself, Alan Turing, who experimented with electrical switching devices as far back as the thirties.

His first primitive "computer" could already solve a complicated equation in about a week, which would have taken top mathematicians six months to work out. In any case, we now have evidence that at least some higher life forms can "think" and reason, just like the latest man-made computers. At the same time, we can observe on a daily basis that a lot of mankind's "reasoning" really is not so superior at all, that it is often based on animal instinct, superstition and illogical feelings.

The reality is, homo sapiens is basically an extremely fragile creature, compared even to most species of the same family of mammals — the primates. We have no fur, no feathers or any other built-in protection from temperature fluctuations. Our prehistoric ancestors HAD to use their hands to make primitive tools and they HAD to use their brains in order to survive, and because the rest of their bodies were no match against other predators and rather useless for catching prey. The more they used their brain, the more their bodies deteriorated at the same time . . . another law of nature to which I will refer to again.

Our eyesight is much poorer than that of a great number of other animals. An eagle can spot a mouse from a distance of one and a half miles. We can barely lift double our own weight, unless we are weight-lifters, while many insects can lift 30 to 300 times their own body-weight. Our sense of smell is greatly inferior to many other members of the family of animals. We just have to look to our pet dogs to verify this statement. This applies to our hearing also, the sensitivity and frequency range of it.

Elephant seals can dive to a depth of over 3000 feet deep and stay there for several minutes, while they feed at the bottom of the ocean. They are true champions of divers and spend 90% of their life in and under water. In comparison, it is considered an accomplishment when a human diver manages to dive to 200 feet without a pressurized suit.

Most mammals can MOVE much faster, on land, in the water or in the air. We have tried to imitate the birds' flight for centuries. Our capacity to swim and to hold our breath under water as long as whales and dolphins is no match. Water fowl can not only fly and dive, they also float on water. They are normally unsinkable. Simple piranha and goldfish have infrared vision, which only the most sophisticated human technology can match. Bats register ultra high frequency sound and are able to navigate by ultrasound.

Dolphins, the "primates of the water," in addition to superior physical capabilities also display compassion, joy and anger just as we do. Recently, wild dolphins have been successfully used for therapy to handicapped children. And here is a good one: the common flea can jump 150 times his own body length and the acceleration for the jump is ten times that of the NASA shuttle after take-off! A person five feet, six inches tall would have to be able to jump about 825 ft. to match the flea!

The hive of the common honey bee is a good example of a complex and well-organized society in the insect world. Biologists are still not clear how the bees decide on their various duties, but once a bee becomes a "worker," the rest of its life is devoted to this duty. Some worker bees are known to specialize as "morticians," cleaning up and disposing of dead bodies within the hive. Others are "guard bees," acting as "immigration" officers by checking every incoming bee and rejecting intruders from other hives. Other workers have the dubious duty to circulate the air near the queen bee by fanning their wings, thus providing "air conditioning." Others are assigned to "scouting" duties by locating new sources of nectar, by reporting to the colony by performing some sort of "dance," as indication of the new-found food supply for the "collector" bees.

Another animal worth mentioning at this point is the weaver bird. This bird is capable of creating a small architectural wonder, a rain- and wind-proof nest of such unique design, that potential intruders cannot enter it. The entrance is facing down to avoid rain from entering and if the nest is not 100% to the male bird's satisfaction, he destroys it himself and starts all over.

Elephants can hear and communicate by "infra sound," extremely low frequency sound waves. Blue whales can communicate over distances in excess of 100 miles, sometimes even as far as 1000 miles. However, man-made pollution of the oceans seems to hamper that capability lately, as scientists discovered in amazement. A desert fox can hear the slightest noise over great distances, even the movement of an insect in the sand. The same applies to the North American kangaroo rat. A barn owl navigates by sound waves in total darkness. These facts were unknown to our biblical ancestors, and can not be found in the Bible, the Koran or in other scriptures.

The point is again, we should not underestimate other living creatures, maybe they are ahead of us in some ways and despite their less developed brain. At least what little brains some of them have, it IS usually used much more to their ADVANTAGE. Unfortunately, we cannot say that about the supposedly most sophisticated brain of them all — the HUMAN brain. As individuals or as societies, we obviously don't always use intelligence to our advantage.

INSECTS AND EVOLUTION

Getting back to insects, I want to add at this point a quick reference to the reality of evolution, which is usually disputed by religious fundamentalists. Insects give us the best opportunity to study the effects of evolution, thanks to their greatly accelerated life cycle. One example are the African and the European honey bees. Their appearance is so similar that biologists cannot detect a difference. However, the much harsher environment in Africa, caused by excessive heat, frequent drought, regular bush fires, etc., caused this species to be much more hostile and aggressive. The African bee also achieved a much higher rate of reproduction, and almost twice the life span of the much more docile European (and North American) honey bee. The evolutionary change took place over a relatively short time of a few thousand years, which translates into possibly millions of generations of bees.

We can pick one other insect community at random. This time I want to refer to the leaf-cutter ant. To begin with, they are capable of transporting objects many times their own weight, similar to many other insects. The leafs are NOT consumed by the ants. They are stored in a special area, where a type of "fermentation" takes place. The resulting fungus growth is the actual food source for this species of ants. Evolution made that method of food production most attractive to this creature.

We may not think in terms of insect "intelligence," but sometimes insect colonies appear to live in greater harmony and with better organized life styles, than our highly advanced human societies. I am not implying that some insect species are more intelligent than man, but some of them ARE using their intelligence more to their advantage. Much more than man does. Most insects have been on Earth hundreds of millions of years longer than modern man, and without trying to exterminate other insect species or other colonies of their own kind. Without nuclear and other weapons, nature has been able to maintain a delicate balance between predators and prey, even in the insect world, as long as man did not interfere.

Getting back to the capabilities of survival techniques, this time of creatures other than insects. The Australian desert frog builds himself a time capsule and survives in a state of suspended animation for up to five or six years, during frequent and lengthy periods of drought. When the rains finally arrive, this frog "eats" himself out of his enclosure and leaves his underground cave, to eat, to mate and to lay eggs. This is the only way for a moisture dependent species to survive the harsh desert environment during long dry periods. A marvelous example of evolution at work -- can mankind match it?

Other desert creatures may have tried to survive many thousands of years ago in this or another manner, but have unfortunately not succeeded. Many of them are now long extinct. Most of man's ancient predecessors have gone the same way, others have been able to adapt their bodies to less hostile surroundings and managed to survive through evolutionary changes. Nature's law of the survival of the "fittest" and the most "adaptable" has been and still is at work for ALL living creatures, with or without religious belief.

The desert's human inhabitants only survive by moving to food and water sources during prolonged droughts. They are unable to survive by hibernation, and prayer does not help. Many of us humans anywhere on Earth have at one time or another wished, we could crawl into a time capsule and hibernate for a number of years and come back, when things are hopefully better!

THE POWER OF DNA AND GENES

DNA, the building blocks of our genes, links all human races, past and present together. In harmony with the modern findings of science, many native people believe in their closeness to nature and with life of all other creatures. They look upon other animals as their brothers and sisters. That was until modern men of many major religions forced their belief unto them, in many instances with devastating results.

The ancient and sometimes quite logical believes and traditions were ridiculed by the early "Christian" settlers in North America and many Indian nations were either completely wiped out in the process or they were badly confused by the introduction of Christian belief. The scientific facts are that DNA and genetic mutations ARE the real source and the end-result of today's life forms, contrary to what religious belief is still teaching us at the end of the twentieth century.

To confirm how much all life is controlled by genes, scientists have recently discovered that daily routine activity, such as sleep, is initiated by new-found genes in mice. It will only be a matter of time when human routine activities can be regulated by manipulating the same genes in humans.

Let us look again at our resistance to diseases, which is generally inferior compared to our closest relatives, the primates and other mammals. Mankind has to have a balanced diet to resist various infections and diseases, or to do one better, we take vitamin and mineral supplements. For example, how long would a human stay healthy on a main diet of raw meat ONLY, as almost all wild members of the cat family do? Or on a main diet of raw fish, the food source for most sea mammals?

An extreme example of a warm-blooded creature living on an unbalanced diet and poisonous to the human digestive system, is the vulture. Vultures DO live mainly off decaying flesh, intestines and skin of other deceased creatures. And they reach a rather ripe old age of twenty years (plus).

Looking at those facts, does it not become rather obvious that ancient Homo Sapiens HAD TO USE THEIR BRAINS in order to survive? Their struggle for survival was much greater than it was for our "fellow" mammals. Man's ancestors started to make covers from animal furs to protect their bodies during the ice age climate. As a result of it, some of our early predecessors survived. However, through evolution and over hundreds of thousands of years, their natural resistance against temperature extremes became increasingly weaker and modern man has almost no protective "fur" left. Together with the deterioration of unused wisdom teeth and small toes.

At the same time, our brains developed further, since we had to use it to an ever-increasing rate to survive. It turned into a vicious circle. When man became less able to outrun prey and predators, he had to invent crude weapons to avoid starvation and extinction. Although, the increased use of the brain by our ancient ancestors allowed Homo Sapiens to survive, other less used body parts weakened at the same time. Fact-based common sense tells us, these phenomena will continue, as long as our species survives on this planet. Again, a mythical "God" and religious faith had nothing to do with it. We can blame it all on evolution.

One of the unwritten laws of nature is that all body parts will deteriorate if they are not used close to capacity, including the brain. It happens relatively fast, as it is known from victims with fractured limbs. On the other hand, our children go to school to develop their brains. Without training through our parents and in schools, the human brain would be just barely more usable than that of many other animals. Just as a child's body grows fastest before sexual maturity, so does the brain have the greatest capability to absorb information or to learn during that age period. Mankind from the most primitive natives to the most technologically advanced societies must continue to "exercise" their brains out of necessity, in order to

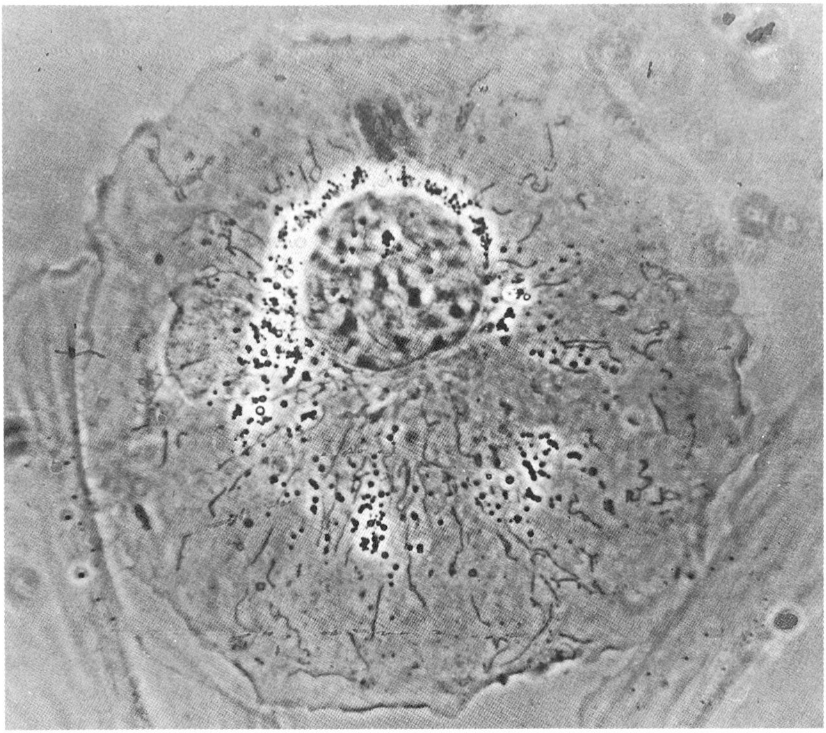

An example of a LIVING CELL — coming in all sizes (usually microscopic) and configurations, they are truly responsible for LIFE and DEATH. From single cell organisms to countless "colonies" of billions of human body cells, along with other microorgasmic life forms, cells contain and transfer good and bad genetic information . . . "making" life by dividing and "breaking" it by dying.

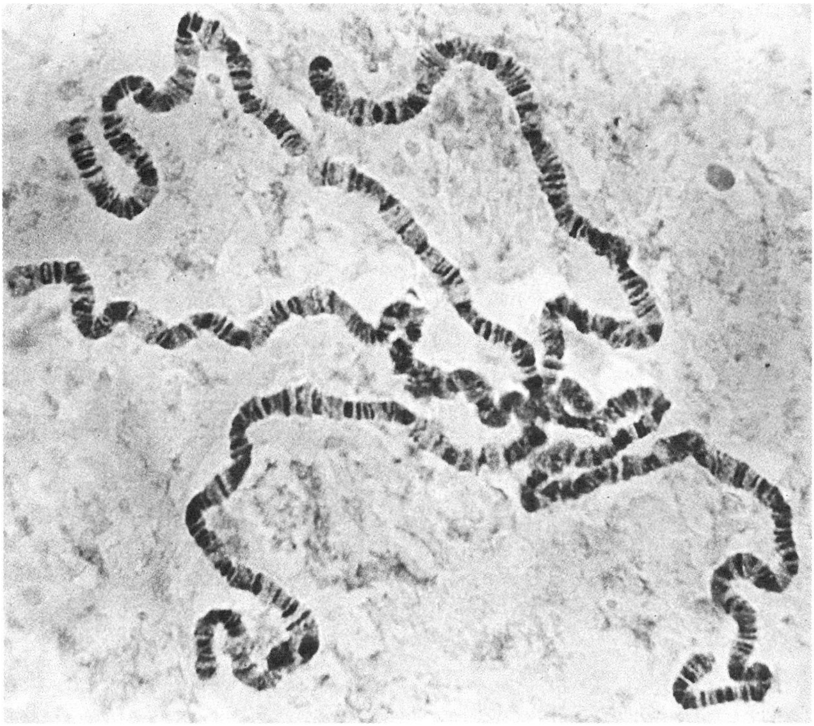

CHROMOSOMES — along with the nucleus the most important part of the cell structure. Their genetic DNA information determines "good" and "bad" hereditary characteristics and mutations within all plant and animal life. Perhaps, THEY should be identified as the REAL "Gods" and (little) "devils" in our life?

survive among most other, physically better developed creatures, or to be ahead of other, potentially more advanced societies.

Good examples are presumably abandoned children, who had to survive in the wild as part of a herd of mammals, often wolf packs. During the 1800's, a young boy was observed living like a wild animal in the forests of France. The boy was captured and initially placed in an asylum. Later, a professor studied his behavior and named him "Victor." He was taught how to play and dress, but he continued to behave much like a chimpanzee. However, his training came somewhat late and he never really advanced to the educational level of an average citizen of the day.

Other seemingly abandoned children have been known to survive in Scotland. A boy living in the wild in Syria was capable of running as fast as 50 m.p.h., and as many as twenty "children of the wild" are known to have survived in the Peruvian jungles during this century alone.

The point is, these children survived according to nature's law of survival of the fittest, without religion, without knowledge of superstition and fear of the unknown, or fear to have committed a religious sin. Their survival was solely the result of a strong natural will for self-preservation and a lot of luck, like any other animal in the natural environment. They could have just as easily been eaten by predatory animals, frozen or starved to death during a severe winter if there was no food available.

Nobody knows how many children have actually NOT survived in the wild and perished as "victims of circumstances." In either case, no mythical "God" ever appeared to have helped any one of these children, the ones that survived or the ones that did not, they all lived or died without religious faith. Their lives were controlled by microorganisms and circumstances, just as ours are, not by faith. Only, we are now able to add common sense and scientific knowledge to "circumstances."

Unfortunately, due to the lack of knowledge, humanity had to grasp at ancient religious teachings. Much time has been, and still is wasted by teaching children and adults one of a multitude of regional mythical religions first, before teaching basic science. Although religious education has been around for centuries, fact-based scientific education was neglected until fairly recently. And little was known about nature and the Universe until the last part of this century. Until the mid-century, astronomers did not realize that our Milky Way galaxy is only one of billions of such star colonies. The "fuzzy patches" in the heavens were thought to be gas nebulae, and not too long before that, mankind did not realize that our sun is only one of billions of other suns, as every star is just another "sun." And not long before that, Earth was thought to be the center of the universe and flat.

Most school children in separate (religious) schools may know a lot about the biblical story of creation, but at the same time they still have difficulties in spelling and in performing simple calculations without a calculator. Where are the priorities? Looking at one typical example: All school boards in Newfoundland, Canada, are controlled by a variety of churches. It is needless to guess, where the priorities are in educating the children of Newfoundland. And what the level of education is aside from the unnecessary high administrative costs to the local taxpayers. Although it is "too little, too late," local governments want to make changes now to this system after receiving continuous pressure from citizens. There are numerous primary and secondary schools throughout the world, where religious education is still an unquestioned priority.

We now have overwhelming evidence that nature applies the same laws everywhere in the universe that we can observe and record today. Theoretically and quite logically, evolution must have the same effect on other planets and in similar solar systems. The evidence supporting this theory is mounting in recent years. Evolution is triggered by changing conditions on stars and planets everywhere and it is never-ending.

Coincidental with the theories of evolution are eyewitness reports of sightings of aliens, describing the visitors' appearance generally to be smaller and physically much weaker than we, without visible body hair but with large brain cavities and big, dark eyes. Their small mouths also make sense, whether one believes in alien visitors or not. I come back to an earlier observation; Whatever is NOT used in nature will deteriorate, whatever is used to full capacity will develop even stronger and better. Just another FACT of nature.

We may have the same appearance in two or three million years from now as alien visitors of today are said to have, even if the climate and environment on Earth do not change significantly during this period. It must be more than coincidence that the aliens' appearance is reported by eyewitnesses to be so similar to a "logical" image, as could be expected. Or could hundreds of self-proclaimed "eye witnesses" have the same vivid, but somehow logical imagination? Can they ALL be wrong or have they all fabricated hoaxes?

Since most mammals or animals in general did not have to use their brains as much as our own human ancestors, their bodies have developed better in many other ways, depending again on the degree of usage of the various body parts. TIME in nature is always one of the most important factors, only the strongest survive or the entire species disappears like the dodo bird.

If the climate or the environment around some species changes too fast for their bodies to adjust, the entire species will become extinct. We have overwhelming evidence of it through the dinosaurs and other entire generations of species, which fell victim of massive, instant natural disasters. Today, with the natural low reproduction rate of elephants and rhinos and without natural enemies, the mass killings by modern "civilized," sometimes religious men of "superior" intelligence, could wipe out those magnificent species in the same way.

Many deeply religious populations of South American and European countries still practice the torturous "sport" of bull fighting. The same people, whose ancestors caused the extinction or near demise of South American native tribes when they invaded their land and forced Christianity upon them. Others eat frog-legs as an unnecessary food delicacy, after these harmless creatures have their legs cut off and are left to die. Many times churchgoers have this "delicacy" for Sunday lunch after church.

Getting back to the elephants and rhinos, these ancient animal species are only two examples of life forms which are unable to adapt to rapid changes in their "environment," (man-made changes in this case), contrary to most insects with a much shorter life span, and the capability of massive reproduction and with higher resistance to environmental changes. After all, most insect species have survived mankind's ongoing war against them, a war fought with powerful insecticides. Instead of exterminating the insects, higher life forms such as birds of prey and man himself have contracted genetic defects and diseases from the chemicals through the food chain or through inhaling the contaminated air.

MORE ON THE REAL BUILDING BLOCKS OF LIFE

At this point, this may be helpful to the reader to explain the meaning of the most important elements for the creation of life and for sustaining it, and not to be found in the scripture:

AMINO ACIDS: large molecules containing carbon, nitrogen and hydrogen. Twenty are present in living tissue as building blocks for protein.

DNA (Deoxyribonucleic Acid): A "double helix" molecule carrying genetic information.

RNA (Ribonucleic Acid): A group of molecules transmitting genetic information from cell generation to cell generation, aiding DNA.

PROTEIN: Complex, chain-like polymer molecules made up from amino acids, essential component of all living cell formations and carrying out the cell's chemical reaction. However, plant protein is incomplete for human consumption, only animal protein has all the amino and other acids required by the human body. Protein is the most complex molecule known to us, made up from more than 2000 atoms, compared to water molecules with only three atoms.

NUCLEIC ACIDS: Another type of polymer molecules, which embody the DNA and RNA genetic code.

CARBOHYDRATES: Carbon, hydrogen and oxygen atoms forming an organic compound.

CATALYSTS: molecules and elements causing acceleration of chemical reactions.

NUCLEOTIDE: building blocks for nucleic acids, mainly carbohydrate, nitrogen and phosphorus compounds.

AN EVER-CHANGING UNIVERSE — NATURE NEVER STANDS STILL

Another very significant unwritten law of nature appears to be "A Consistency in change," mainly evolving around environment. All life is shaped around, and influenced constantly by climatic changes, atmospheric changes, in solar radiation, etc., and by time itself. As noted before, nothing remains stationary in nature or in the universe in general, everything from the tiny atom to the entire cosmos is subject to evolutionary changes in one way or another, caused by a never-ending chain of events. Just as our bodies' physical appearance will not at all be the same in 400,000 years from now, all other animal species will have changed with few exceptions, such as the cockroach and the horseshoe crab, (maybe?).

This statement is based on the facts of the past, since only 400,000 years ago most mammals appeared different from today's version of the same species. Mammals' bodies, and particularly primates' bodies, seem to have to change more noticeably to adapt to the ever-changing environment of the surroundings. They are simply more delicate and do not have the obviously more resistant bodies of crocodiles, sharks, right down to cockroaches and other insects. Some of them do not have the capability to multiply in great numbers. Crocodiles, sharks and roaches may also not have changed their physical appearance during tens of millions of years maybe based on the principle: Why change a good thing?

Roaches are a good example in any case, their bodies are so tough that they will survive, and have in the past survived the worst conditions on Earth. That didn't do much for their brain, when a creature can survive and multiply in the worst possible environment, when it can go without food for unbelievably long periods of time and when it can feed on almost anything, the brain is hardly ever needed. The brain does not have to develop any further over millions of generations and what little there is, is obviously most sufficient for survival. Once again, we see why the human brain HAD to be used all along in order for our species to survive.

Dinosaurs were not so lucky. The combination of their small brain AND their large body size may have been their downfall 60 million years or so ago. Only the smallest and the more intelligent creatures were able to "hibernate" through the months and years after the massive natural catastrophe of that time.

Scientists are now admiring the common horseshoe crab. With a near-perfect body for the environment in the oceans and with little change to ocean "climates," compared to the drastic changes to the planet's surface the horseshoe crab did not have to adapt to anything new. Nor did it have to use its "brain," but it still survived more than 300 million years. Its uniqueness allowed this creature to survive against all odds. Other species have become so highly specialized that only a very narrow change in environment causes havoc or extinction. However, to add a sad note to the long history of the horseshoe crab, human overharvesting in recent decades could prove dangerous to the survival of this astonishingly tough species.

Many animal species have survived simply by devouring their environment. The waste of one creature became food for another. A perfect recycling system, created by nature. If one partner of this "recycle family" becomes extinct, the other partner species will soon follow. This statement not only applies to animal waste and the bacteria living off it, but it also applies to all larger predators and prey.

On a closely related subject, laboratory measurements with lie detectors have been successful in proving "primary perception" of the most primitive life forms from bacteria to plant life. In other words, bacteria have the capability to find "food" sources, even though they have no brain or "senses" of taste or smell. Plants have reacted even to sound, with negative results from loud hard rock music and positive reaction from mellow classical music. All plants were subject to identical conditions otherwise (temperature, moisture, light, soil etc.), while the one lot of plants deteriorated and finally died after being close to the loud hard rock, the other lot thrived on soothing classical music.

Obviously, if the plant's primary perception is capable of reacting to various sounds, there is no doubt that their roots for example can "find" the best areas of fertile soil in the immediate neighborhood by growing toward those points. It is easy to observe plants growing in the direction of light by simply turning house plants around. The point of this little demonstration is that all life forms have adapted to their own environment through more than three billion years of evolutionary processes.

Mankind is no exception, had our very early "ancestors" decided to remain aquatic or amphibious, we would look more like dolphins or reptiles today. It is the environment and evolution, which shapes all life on this planet, NOT a hypothetical, mythical "God" chosen by many to be the "divine creator" of all life. The proof is overwhelming.

Modern man consistently upsets the natural recycling procedure, by "controlling" certain animal species which happen to be unlucky enough to have to share our habitat. Wolves and bears are only two examples. First, we killed so many wolves that they became an endangered species, then, when their prey (deer etc.) became over abundant, we started to "control" their numbers.

Getting back to the horseshoe crab, it has been only in the past few years that scientists have processed the crab's blood into a product called Limulus Amoebocyte Lysate. This is used to detect "endotoxins" (toxic substances in certain disease-producing bacteria, liberated by the disintegration of the bacterial cell) in human diseases, pharmaceutical drugs and even in medical devices such as artificial hearts. No wonder, this crab family has survived hundreds of millions of years through unbelievable hardship in nature. Medical science will no doubt discover more surprising remedies from the body fluids of this, most likely the oldest surviving unchanged animal species.

LIFE EXISTS WHERE IT WAS NEVER EXPECTED — AND — THE POSSIBILITIES OF A "BIBLICAL FLOOD"

Creatures living in extremely hostile conditions at one time or another had to adapt their body functions to the specific surroundings, or they would have never survived. Deep Sea life has managed to thrive and multiply for millions of years in total darkness and under tremendous water pressure 10,000 or 12,000 feet below the ocean's surface. Any mammal, including human beings which have all evolved on or near the surface of the Earth, would be crushed within seconds at this depth. The reverse applies of course. Deep sea creatures could never survive near the surface.

There is definite evidence of life in the harshest parts of the world's desert areas and under the arctic ice. I want to emphasize again, life as we see it around us and what we can see a few meters under the surface of lakes and oceans, has evolved by changing with the environmental surroundings over a period of several hundred million years. The evidence is simply overwhelming and it was not known to humanity 3000 years ago, during a time period, when religious faith and scriptures had the only "solutions" to the creation of life. During this time nobody knew about life 10,000 feet below sea level and under polar ice floes. Mankind did not even KNOW about the existence of polar ice caps.

The importance of microorganisms to human life and the origin of life in the oceans 500,000,000 years ago were equally unknown 3000 years ago. That brings me to the biblical belief in Noah, who is said to have saved all creatures by placing a pair of each in his Arc. What about the deep sea fish, or without going down that deep, how did he transport a pair of polar bears from the arctic and penguins from the antarctic regions, as well as hundreds of millions of insect species from all around the world?

Religions have not taught us these facts, but modern science has. Logic tells us therefore, since all life forms have changed in the past 100,000,000 years alone, there is no reason to believe that life on Earth will NOT continue to change in the next 100,000,000 years also. If we could come back after that time period, we would probably not recognize most of the creatures and plant life, including humanity. Homo sapiens is just another mammal that has and will continue to change along with the rest of them.

Much depends on our planet and our ultimate source of life, the SUN. A hypothetical slight, but rapid temperature change from our life-giving star would have overwhelming and eventually fatal effects on most life-forms on or near the surface of the Earth and the oceans. Again, maybe with the exception of cockroaches and their "relatives." But even if there are no major changes coming to the Earth's environment, or ultimately, from the sun, all life forms will continue to try to adapt better to the prevailing conditions. However, we can relax. Our Sun WILL continue to shine at about the same strength for at least another three to four billion years. After that, things could become increasingly uncomfortable, and the religious fundamentalists WILL have their day through the long awaited "Armageddon." However, it will be caused by a rapidly expanding sun. It is predictable and not the wish of a hypothetical "God," but more about that in the Astronomy section.

Over the next hundreds of millions of years, our brains will continue to develop more, the amount of body hair will decrease or completely disappear. Our teeth will become shorter lived and weaker, along with a number of body parts, such as the small toes. Evolution will affect other mammals in the same way, always based on the principle: "whatever body part is not used to near capacity will deteriorate, whatever part is used and exercised to over capacity will further develop."

We have proof of what happens even over a short number of years, when we examine active athletes. Their hearts become larger, along with other heavily exercised muscle tissue. One could just imagine what 10,000 generations of athletes would look like, not to forget how chromosomes, genes and mutations would cause that strain of human beings to change over a few hundreds of thousands of years! Just as generations of bees have changed in compliance with this law of evolution. We therefore have "worker" bees, "army" bees, "queen" bees, etc., their appearance has changed in accordance with their PREDOMINANT physical activities.

As I already stated, the slightest change in solar radiation intensity alone could turn the sun from "life giving and supporting" to "life taking," or at least it would cause major changes in the evolutionary chain of events in the entire solar system. Science now can forecast such changes, originating from our sun or from other extraterrestrial sources. We have a good example from modern-day scientific observations of the ozone layer in the stratosphere.

Only a few hundred years ago, if a scientist had predicted changes to the then unknown ozone layer, caused by either man-made or by extraterrestrial sources, nobody would have believed him and he would have been the laughing stock of the day. It may have resulted in severe punishment, after a religious "witch hunt." If the changes did not agree with religious 2000-year-old prophecies, this individual would have been doomed and may have ended up on the scaffold.

Once again, we must seriously look at the dates WHEN the holy scripts were written, and by WHOM! Theologians are not sure themselves, WHO the authors of the Bible were.

As a matter of fact, NO-ONE really knows who the authors of the Bible were. In today's terms it may have been an excellent science fiction book at the time, since few of the contents are based on scientific reality or logic. It is also said that the Bible is the "word of God," yet there is no evidence that the mythical God wrote it.

The various versions of the Bible are the world's biggest best-sellers, teaching humanity little of the realities of the twentieth century, only a lot of wishful thinking and fiction. Despite today's scientific knowledge, so many still accept the Bible and other mythical religious books as

"FACT" and do not dare to question its validity and lack of common sense. Yet, at the same time mostly the same people and certain government leaders question scientific discoveries and logical theories. They still apply the philosophy; those who do not believe in the same religion as I do are evil and must be eradicated.

Today we are privileged to live in an age, when the greatest number of discoveries is made, giving us a choice in priorities between traditional religious BELIEF and scientific FACT and logic. And we can back up only scientific discoveries with facts.

Scientists do have scientific evidence about the occurrence of a "biblical flood," with its devastating effects on Earth. 3000 to 4000 years ago, when biblical history was recorded, the TRUE scientific reasons of creation and nature were unknown to humanity of the day. It was only natural therefore, to write it off as "God's will" for punishment of earthly "sinners." In reality, we have strong scientific evidence that an asteroid or comet crashed into Earth, or at least caused a major temporary change in our planet's orbital movements. In any case, an event of this magnitude would alter the planet's magnetic field and possibly, the speed of rotation, a perfectly normal natural phenomena. It happens all the time in the universe and by chance.

Only NOW are scientists able to forecast many of such catastrophic events, saving possibly millions of lives. The probabilities of a similar natural catastrophe to occur again in the near future are slim and can be based on nature's unwritten "law of probability." Since most cataclysmic events are now scientifically predictable, the "will" of a superhuman "Creator" as a source can be eliminated.

Without going into detail now, (that is for the chapters on astronomy), astronomers discover almost daily more evidence that the entire universe is made up of the same atomic matter. By now they know as a fact, all matter in the detectable universe is subject to the same laws of physics and chemistry as our own sun and our own planet, only at different stages of development. Whatever happens to Earth at one time or another, can and will happen to other planets, regardless if they are inhabited by "human sinners" or not at all.

Although the proportions of atomic matter differ from one star system to another, we know exactly why our sun generates as much heat as it does, and how long and why larger or smaller stars react differently. It is based on mass, age and location in a galaxy, amongst other factors, but more on that also in the chapters on astronomy.

The following is quite conceivable: shortly before an asteroid was either going to crash on Earth, or come dangerously close to its gravitational field, (just before the "biblical flood"), someone, (Noah, through a crude "telescope?") had the intelligence to observe an extraterrestrial object heading for Earth. Few people would have believed his prophecy had he NOT given "God" as the real architect for this impending disaster. In any case, scientists continue to find more scientific evidence of this and other major terrestrial disasters of the distant past.

To illustrate the possible cause of the biblical flood in more detail, we all know that our planet rotates once every 24 hours at a precise and constant speed. The gravitational force of a large extraterrestrial object, such as an asteroid of a few miles' diameter, passing close to Earth or colliding with it would temporarily slow down Earth rotation by a fraction. All water, air and loose surface objects would continue on at the same speed, just as during an automobile collision when it takes seven-tenths of a second for everything to come to a complete stop. It is all part of the laws of gravity. Or when we swing a bucket of water around at a fairly steady speed and bring it abruptly to a halt. Every child knows from experience that the water keeps on going for a few seconds! This experiment also confirms the law of gravity.

Slowing down the Earth's rotation just barely, and only over a very short time period would cause devastating floods, storms and surface movements. As during the "biblical" deluge, water would rush up and over mountains.

Since the Earth's rotation is from west to east, the western coast lines of all continents would be completely submerged first by the in-rushing ocean waters. The east coast of a large continent would display the opposite effect and all water would be temporarily rushing away from shore, only to return shortly thereafter causing devastation until Earth's normal rotational speed is restored.

This phenomenon happens at all times in the universe, collisions of much greater magnitude are common and are simply based on the laws of physics and on the law of averages, nature's law of probability. As a result, with the vast numbers of "bodies" in the universe, such collisions are unavoidable, even in the vastness of space and particularly within "young" solar systems. Just as accidents happen on highways at any time, even during periods of minimum traffic. In any case, evidence of frequent collisions between objects in the universe is simply overwhelming.

A hypothetical repeat of the so-called biblical flood would not necessarily cause extermination of ALL life on Earth. After all, some of the ancestors of today's life forms on our planet obviously survived previous disasters. Remember the horseshoe crab? Anyway, I have to think about it again, how could they possibly ALL have fitted into Noah's arc, a pair of each of hundreds of thousands of living creatures?

No, we will not have to build another Noah's Arc. Depending on the severity of a hypothetical, but probable, future repeat performance, shelters on Earth may suffice. Or at the very worst, a number of nuclear explosive devices launched from space stations and strategically exploded near the intruding object, would deflect it and save millions of animals and people. This would be one time when all future scientists would have to work overtime to SAVE lives and NOT to destroy them. The important point is that we have the capability today to prepare for predictable natural disasters through advanced science only available in the twentieth century.

Also, no matter how devastating a natural disaster may be, some life forms will seemingly always survive. As a good example, scientists have only now discovered that primitive bacteria have not only survived the massive Mt. St. Helen's eruption, these micro organisms are actually thriving in temperatures of more than 100 degrees Centigrade, (above the boiling point of water). Similar cultures of bacteria have been discovered near underwater volcanoes where the water is hot enough to melt lead, where water meets the hot lava. Laboratory tests are under way now to use this type of bacteria in the fight against certain types of pollution. Similar cultures of bacteria are already used experimentally for the cleanup of oil spills since certain bacteria actually thrive on crude oil! Nature has an antidote for every evil, we only have to discover it.

A logical conclusion therefore is, the original life forms on Earth about three and a half to four billion years ago, may have been the same bacteria-type species living under similar conditions. When we come to think of it, we boil vegetables and other organic food, and not ALL micro organisms are destroyed through this procedure. Another fact of nature which has been right under our noses, but we have not paid attention to it, until now.

Our planet and the entire solar system have been, and will be targets for extraterrestrial debris of all shapes and sizes again and again, based on the law of averages and on the history of Earth. Anyone with a pair of binoculars can easily observe the results on the surface of the moon, where countless collisions left craters of all dimensions. Without doubt, Earth ended up with the same amount of impact craters mainly during prehistoric times, only atmospheric conditions on this planet erased most impact craters. Just to name two obvious ones, the approximately 50,000-year-old, 600 feet deep and 4000 feet wide, meteor crater in Arizona and a much larger and older crater in northern Quebec, Canada, Lake Manicouagan with a diameter of about forty miles.

A mysterious and massive explosion on June 30, 1908 in the remote Siberian region of Tunguska had to be also of an extraterrestrial origin, as I will explain: The blast from this explosion was so powerful, it could be felt by inhabitants 200 miles away. An uninhabited area of 1500 square miles of forest was devastated by the tremendous explosion. The power of destruction was equivalent to that of an extra large H-bomb, or by conservative estimates, at least ten times as powerful as the atomic bomb of Hiroshima. Yet, it left NO CRATER.

All trees had either been flattened or snapped into pieces, or both, over the entire area. Near the center of the explosion, all tree-trunks were pointing away from this area like the spokes of a wheel. Many years later, (former) Soviet scientists detected radio active material in the rings of some of the trees, and alien matter was found in the mosses of local swamps, which have never been recorded on Earth, then or now. It has only been discovered in recent years how much similarity the devastation of the Siberian forest had with the A-bomb blasts of Hiroshima and Nagasaki: no craters, and trees flattened and left in a typical arrangement of the spokes of a wheel pointing away from a central point.

The stories passed on from inhabitants of the area also coincide with the description of a nuclear explosion, and in no way with the crashing of a comet or a meteorite. The shock wave around the Earth was registered twice, and many countries around the world had strangely colored, spectacular sunsets and sunrises, presumably from fallout clouds. Yet, humanity was extremely fortunate that this devastating explosion did not occur over a populated part of the world; literally millions could have been killed. But chances for that were about as remote as winning big in the lottery or being hit by lightening in the middle of the northern winter. Religious believers may say that "God" was not ready yet to punish the sinners of the planet, at that time.

Most scientists strongly believe or accept it as fact that it was a nuclear accident originating from extraterrestrial sources. In reality, a nuclear explosion, OTHER than one caused by intelligent beings is most unlikely and is not known to have ever occurred. The evidence therefore points strongly toward an accidental explosion of an alien nuclear powered craft, since humanity had no knowledge of atomic power or how to create a nuclear explosive device in 1908.

Other astronomers believe, it was part of a comet which exploded before hitting the ground. It is also a somewhat acceptable theory, but surviving eyewitnesses did not see anything that could remotely resemble a comet. The lack of a crater makes it difficult to accept a meteor as the cause of the explosion. The point is that our planet and all other celestial bodies are frequently "victims" of such unpleasant events, with or without impact craters.

It also makes no difference if the planet is inhabited by people or not, and if Earth is targeted again, it is without consequence of faith of the inhabitants of the impact area. Small meteors have already hit populated areas, as well as remote locations. Of course, with most of the Earth uninhabited or covered by water, chances are much slimmer for a city to be hit. No death or injury has ever been recorded in recent history as a result of a meteor impact.

These impacts take place due to natural events which may be expected, and not because some of the inhabitants of our planet have been "sinners" or believers in the "wrong" religion or that they are being punished by an almighty superhuman creator. Of course, the same applies to volcanic eruptions and earthquakes here and on other known planets.

Facts and logic give us the answers. Just as in "Noah's" days, when someone had primitive scientific foresight and was able to observe an extraterrestrial body heading for Earth. Isn't it reassuring to know that we can actually forecast such an event to the day and to the hour and with a good idea of where on Earth the greatest damage will be caused? Thanks to twentieth century science. Contrary to it, religious scripts only confuse humanity with their "prophecies" of doom and gloom.

In this respect, many of our fellow mammals seem to have another advantage over us, they appear to have an extra sense to foresee danger. It is a known fact, that many pets or domestic animals will get very nervous and restless just before an earthquake. But no "God" has warned us of the precise timing of coming volcanic eruptions or of major earthquakes, most definitely not with a precise forecast of the place, date and time of the expected impact. Prophecies of "Armageddon" have fortunately so far been incorrect, although some fanatic believers not only prepared for it, some even committed suicide in anticipation of the prophecy. It's amazing what a strong belief can do to the human mind.

ONGOING BATTLES: INTELLIGENT REASONING AGAINST SENSELESS BELIEFS AND EMOTIONS

A number of people have reported seeing Christ, God, the Virgin Mary, Satan or some kind of "demons" or "angels." Interestingly enough, there was never any scientific or physical evidence left behind. The Virgin Mary or Christ was only sighted in countries with a Christian majority, seldom by believers or nonbeliever in predominantly Islamic or Buddhist countries for example. The "sightings" were usually reported by one individual only with few, or no witnesses present.

Occasionally, Priests and other Christian clergymen claim to have removed the "devil" or "demons" from various persons through "successful" exorcism. One powerful mind controlling another through hypnosis maybe, fact — NO. As a believer's saying goes, those who believe need no proof — those who do not, will never find proof. I believe we WILL find it.

Millions including myself would rather believe in reality and accept proven facts, the existence of "angels" or "demons" does not fall into this category.

It appears quite noticeable that visions of angels (or demons) are, as a rule, hallucinations induced by drug use, periods of high stress, being overly tired or by other mind-influencing events in the subjects' life. Believing in angels may actually help a strong believer occasionally. However, in the end, no matter how much the believer relied on a "guardian" angel, he WILL die — and in the same variety of ways as nonbelievers.

Believers may also look upon "faith healing" as "fact," as well as sightings of the Virgin Mary and Jesus. As mentioned elsewhere, the human mind can be extremely powerful. When we want to believe something bad enough, we can really "believe" in "miracles," such as faith healing or sightings of religious mythical figures.

I can compare this phenomenon with a headache, just to display one example. Everybody has experienced the occasional headache and has taken an Aspirin, or some other medication, hopefully curing it. Shortly after, the headache is gone. Was it really the Aspirin or was it our FAITH in the drug, what made the headache disappear?

Occasionally, I have full intentions to take some sort of medication against a persistent headache, but I forget or there isn't anything in the house. A short time later, the pain is gone, without medication! Had I taken a painkiller, I would have credited IT with the disappearance of the pain. I would have "believed" in the healing power of the medication. Just as other believers "saw" the Virgin Mary, etc. . .

Medical doctors are frequently testing patients in groups. Half of the group receives a regular medication and the other half a placebo. Many times, the patient's faith in the healing power of the (placebo) medication actually works and the patient is cured. Placebo operations had similar results, it appeared simply because both the patients and the surgeons had a strong desire for success. Volunteer patients who had little to lose through an operation were cured, although the actual operation could not be credited with it. A bit of an ironic twist to the phrase, "the operation was successful, the patient died." In this case, the patient lived and the operation was unsuccessful — faith healing in a true sense of the word.

One could easily consider homeopathic treatment a

"placebo" treatment, since the actual medical ingredients of homeopathic "medicine" are extremely minute in quantity — for all practical purposes, they are nonexistent. Yet, homeopathy works for many patients — is it another example of strong willpower overcoming a disease? Simply put, the patient and the medical specialist both WANT IT TO WORK?

The strong belief of a sick person in fact cures certain illnesses, NOT by the hand of a "Faith healer," or through "sightings" of the Virgin Mary. I emphasize curing of "certain" illnesses, since we have yet to see a faith healer being able to re-grow a missing limb or any other body part, such as a surgically removed kidney or a lung! If THAT would happen, even I would believe in religious miracles!

The "Miracle of Fatima" is another example of a strong BELIEF ONLY. Although many observers came prepared to see and hear the Virgin Mary, only one young girl claimed to have actually seen and spoken with her. Everybody else just saw the setting sun emitting "strange" colored rays. The "light rays" moved toward the observers, as reported in 1917, but their belief may not have been strong enough to see what they came for.

One point comes to my mind in this respect: I have watched movie excerpts from this and other "miracle" sites and found the absence of photographic evidence of the "appearance" of the Virgin Mary or of the "Lady of Fatima" very conspicuous. The footage of hundreds of faces of awe-inspired believers was dramatic, but there was not one frame taken of the miracle "lady." Would it not be the most obvious duty of the photographer to film as much as possible of this "wonder" and only as secondary option, the faces of the observers? After all, other photographers [amateurs usually] managed to photograph images of ghosts and of UFOs!

One mass sighting of the "Virgin Mary" was reported from Egypt in 1969. It appears that this "sighting" may have been based on mass hypnosis turned into mass hysteria. Unfortunately, I found very little information on this event, no photographic evidence, tape recordings of her voice, what language she spoke and who the observers were in this mostly Islamic nation. Did she introduce herself as the "Virgin Mary" or was it all a fraud?

The only logical explanation to many for these events is the human mind and the power of strong, often blind faith. It could be seen very clearly during a TV talk show recently where some individuals, who had claimed to have been possessed at one time or another all had experienced some traumatic event in their earlier life. One girl had been severely abused, physically, mentally and sexually at a very young age. Other persons who claimed to have been possessed, appeared to me to be much more receptive than others to spiritual events. One lady claimed to have lived through a nightmare, but she was convinced it was not a dream. The point is, if a person WANTS to believe in illogical, so-called spiritual events, that individual is eventually convinced of what they saw or dreamed was REAL.

Again, why are usually only single individuals involved, never a large group of creditable and unbiased witnesses? Where is the evidence? Drug addicts are known to have hallucinations of demons and only after the drug has worn off, do they realize that it was a drug-induced dream. Or they forget all about it. The human brain CAN play tricks with us. We have all done something at one time or another, after which we asked ourselves: "why did I do THAT?" Or "What did I think when I did THAT, it was a stupid thing to do." Addictions, such as drugs, alcohol, gambling, tobacco smoking etc., fall into this category.

I want to go a step further and add fundamentalist religious belief and political fanaticism to the list as two more "near-addictions." An addiction and the strong "feeling" of not being able to survive or to live without it seems to reverse the role of the human mind. At that point, the mind is no longer in control, individuals start to act somewhat like robots. They are like an alcoholic, who does not realize anymore when and how much he or she drinks, a smoker who lights a cigarette without realizing it, the gambler who continues to gamble without really being aware of it. It is as if the "wrong" part of the brain has taken over temporarily, human beings doing something at a given time, without a plan or without logic. When this happens, humans seem to lose their imagined superiority again, over the rest of the animal world.

Hypnosis is another good example of the power of the human mind. Only those who believe in it can be hypnotized. Also those who believe in evil spirits and demons or claim to be possessed by one, can in turn expect to be helped through exorcism or "de-hexing" to have the imaginary demon removed. Unfortunately, some victims believed so strongly in demons that they accepted it as fact and in the end died exactly as their demon instructed them to.

Still on the subject of the power (or weakness) of the human mind: Once again, we can consider ourselves privileged to be alive at a period in history with the greatest technological advances, greatest discoveries and with most significant inventions. At the same time, we can now openly question what scriptures taught our ancestors and what we are still led to believe. We can back up our doubts and questions with available and ever-increasing scientific knowledge and common sense. Today, we know that brain functions stop after biological death, and we have no evidence of the power of the mind working after death, contrary to what we learn from religion.

On the negative side, during the past nearly 100 years, human ingenuity has brought more violent deaths and destruction to themselves and to other inhabitants of our planet than during any other periods of time in history. This is because of technological advances, and often inspired by religious differences. Our supposedly not-so-intelligent "fellow mammals" still mostly kill for food, in battle over territory or during mating season over possible rivals. They do not kill over religious differences, or to enforce ancient, illogical traditions.

Although we are led to believe by religious scripts that we are superior to all other living creatures, humanity still fights for the same basic reasons as other animals do. Only we have invented better means of destruction than other creatures with or without brain power close to ours! And we have added different reasons for destroying each other. One of them is difference in religious belief, making mythical religion in the twentieth century a perfectly legal and acceptable reason for "holy wars," state terrorism and destruction.

However, humanity does not kill quite as often any longer for food, we raise it, breed it and then kill it. But we DO still fight over territory, just like most other animals. We prefer to kill each other, rather than to recognize the fact that we are TOGETHER on this planet through circumstances, NOT because our ancestors practiced the "right" religion. No matter what race we are part of, or what religion we have chosen or inherited, we are really not much different from each other. And what physical deviations there are, being largely due to climatic differences, altitude, environment and all part of the evolutionary process through which ALL living creatures have to go.

We are all in the same "spaceship," a large human colony, but only as a tiny part of much larger, ever-changing natural life cycle. We are all subject to the same earthly weather and to environmental changes, to the same natural disasters and sicknesses. The HIV virus and other killer viruses and bacteria infect ALL people, ALSO regardless of religious belief or racial heritage.

As fast as medical science is catching up to eliminate certain infectious diseases, new mutations of viruses appear. The many varieties of flu strains are good examples. And tuberculosis has returned recently, apparently immune to the treatment of past years. The newly discovered streptococcus "A" virus is another example. This virus actually destroys flesh, and the (fortunately) very few victims really had no chance.

One question comes to my mind: Have these "new" viruses been around all the time, just like the reappearing ones (tuberculosis), and we are slowly losing a formerly (inherited?) natural resistance to them? Perhaps man-made pollutants contribute to the loss of resistance?

There is no doubt that medical science will overcome these setbacks also. My point again is that our lives are almost exclusively controlled by MICRO ORGANISMS, and WITHOUT EVIDENCE of a "heavenly" influence, one way or another.

LOOKING FOR COMMON SENSE IN RELIGION

It is written in the Bible: " Thou shalt not kill!" and "thou shalt not steal." Both commandments make a lot of sense and both should be prime motivators in our lives, for religious believers AND for nonbelievers. They are simply LOGICAL. We do not need religion for this moral and any other logical rule of life. Adherence to this teaching and others, if based on logic, is for all of mankind, religious or not. Of course, all biblical commandments which are based on LOGIC, fall into the same category.

Another one of few logical religious teachings is (words to that effect): "Do unto others as they would to do unto you." A negative version of it would be just as valid: " Do NOT do unto others, what you would NOT want to be done unto you." Using this biblical quotation AND common sense, why is humanity still waging wars and armed conflicts in the name of conflicting RELIGIOUS FAITH? Where are the priorities, is it more important to kill fellow human beings because of their difference in religious belief — or should it not be a priority to HELP other human beings, REGARDLESS of their belief? When is humanity at last really going to start using the tremendous hidden power of the mind?

It becomes most absurd, when so-called deeply religious Islamic governments, speak of a "Holy war" against followers of other religions. Or when Christian church leaders support military conflicts and bless weapons of destruction, which will inflict death and suffering to fellow human beings. "Praise the Lord and pass the ammunition." And it is still happening now, at the end of the twentieth century! To this day we can see religious wars causing revenge action resulting in an endless cycle.

As a good example, in 1979 during one of my many business trips to overseas destinations, I met a Christian bartender in a hotel inside the Moslem part of Beirut. While talking to the patrons, his only goal in life appeared to be his determination to kill a Muslim, in revenge for the brutal killing of his brother by a radical follower of the Islamic faith. When he got his chance, (and I am afraid he did sooner or later), it is quite conceivable that one of the relatives of the Muslim victim later wanted to kill another Christian in revenge, and so on.

Still in Beirut, I also had the opportunity to have lunch in a Moslem home, where the host expressed understandable hatred toward the Syrian peace-keeping forces, which at that time were occupying Lebanon. While driving to a movie theater with his young wife, he was stopped by a Syrian army patrol. Nothing special about that, the Syrian peace-keepers were often checking cars for illegal guns and weapons. At the same time, the occupants of the vehicle were questioned about their destination and so on. My host's bitterness resulted from what followed normal procedure: An intoxicated Syrian soldier threatened his wife and my host, with the intention to rape his pregnant wife at gunpoint.

Fortunately, my Lebanese acquaintance had a hidden hand gun in the vehicle and he was able to shoot and kill the attacker at point blank range in an understandable rage. The moral of this story is: Had there been no hostilities and religious civil war in Lebanon, this episode would never have taken place. The numbers of victims from that war were never clear, but I have read about possibly as many as 150,000 lives lost, Christian and Moslem, all in the name of conflicting religions.

I can add one of two visits to Belfast, Northern Ireland, to this episode when my very apologetic Catholic host took me through the city by car. He explained the boarded-up and often burnt-out, abandoned buildings as a great restructuring plan for "former slum areas." The next day, his Protestant supervisor took me in the same car through the same parts of Belfast. His explanation for the condition of the same buildings was more down to reality; the houses had been damaged or gutted by fire and were abandoned during fighting between Protestants and Catholics. And the revenge destruction and killings in the name of religion in Northern Ireland have continued on to this day.

Whenever Catholics were murdered by Protestants in Northern Ireland, the families of the victims did, more often than not, kill in return out of revenge. As a result of it, we just have to switch the identity of the "players" of this deadly game of religious revenge killings, as the vicious circle continues. We can also substitute "Catholics" and "Protestants" with "Israelis" and (mainly Moslem) "Palestinians", and we have the same situation.

In Algeria, religious fanatics and the only moderately Islamic government are engaged in an undeclared civil war. According to unconfirmed sources over 30,000 Algerians from both sides have been killed so far and thousands are permanently injured. In Pakistan and Afghanistan the situation is quite similar, in both cases blind religious fanaticism has taken thousands of lives so far. And the conflicts go on, all in the name of religion. All the believers had learned from religion was hatred for other fellow human beings. They are not "Thinking before Believing!"

Or we can take this last statement to India and insert "Hindus" and "Muslims" into the opening sentence. Surely, those of us who strongly believe in a "God" as a figure of superior power should realize that The "Almighty" would simply not allow mass-killings to take place in his name! His superior "Powers" would make it impossible. After all, is He not supposedly more powerful than anybody else, including Satan? — But I guess, I am trying to think too "logical" again.

To illustrate typical hypocrisy on the part of religious believers in one of the most religious Islamic nations, I want to refer to a little episode during a business trip to Kuwait in 1979. This one-week visit was only one of about twelve different countries in Asia, the Middle East, South America and Europe, where I had stopped during a rather extensive business trip. The well-known Arab hospitality was outstanding as usual in this small Islamic Emirate. My hosts invited me to one of their usual "weekend parties" on an average Thursday evening.

Although I did not really know what to expect, I was prepared for lots of the usual tea and cardamom-spiced Arabic (Turkish) coffee, together with roast lamb, rice and fruit. But I was only partly right with my expectations. A group of men and women met in one of many small desert weekend cement block "cottages". To my amazement, the unveiled women were NOT the wives of the men present, they were somebody else's spouses. Tea and coffee were absent at first, but a supply of Scotch Whisky was plentiful. The walls were decorated with Playboy centerfolds, another absolute No-No under Islamic law, along with the smuggled booze and the extra-marital "gatherings."

Feeling rather uncomfortable amongst so many illegal activities, I inquired, "What happens if the police raids this party?" I was quickly assured: "They are too busy doing the same things we are doing!" The liquor was purchased from foreign sailors by tying up pleasure boats to freighters and tankers anchored in the Gulf. Apparently, the local police force bought their supply from the same sources.

For appearance sake however, every once in a blue moon confiscations of illegal alcoholic beverages actually took place in the seventies, but only after most of the potential buyers were warned first! - The moral of this little story is: Why not abolish most of the strict, illogical religious laws, if nobody obeys them anyway? Of course, this question applies to our western society also, starting with so-called victim-less crimes, such as consuming alcoholic beverages in "dry" counties, abortions and prostitution in other areas, etc.

At the present time it is no secret that during the recent past, the greatest unrest and the most vicious mass-killings are committed under strong religious, so-called "democratic" regimes. Several Middle Eastern countries and the former Yugoslavia are the most obvious examples. Had religion not played such a dominant role in these and other nations, the entire population could really at least co-exist in relative harmony.

It is quite apparent, Palestinian Arabs originate from the same part of the world as the Israelis. What is also obvious, no nation can be successfully governed continuously, motivated by religious tradition, and without little or no application of logic and consideration for common sense and reality of today. In these nations, there is little or no consideration for the REAL origin of all fellow humans. Men and women are not born as Christians, Jews, Muslims, Hindus or Buddhists, they are born as HOMO SAPIENS, one of many primates and with only slightly different appearances between races.

As one example of the application of logic and common sense, a large part of Europe was at one time forcefully occupied by Romans for hundreds of years. Do the residents of that area now hate all Italians, since the Romans were their ancestors? Of course not, no more than the Mongolian peoples of today are hated in Europe because their ancestors, lead by Genghis Khan, murdered, tortured, raped and destroyed European lives and property hundreds of years ago. It is the only logical attitude, we cannot carry a grudge forever as it appears to be the case between many religious fundamentalists. If only "Christian" Serbs and "Moslem" Bosnians could think that way.

The people of Manchuria have just lately overcome their understandable hatred for the former occupying Japanese. They have come to the logical conclusion that continuous hatred benefits nobody. Not so in Armenia and Aserbaijan, or for that matter in Israel, where many Israelis still consider ALL or most Palestinians to be terrorists. Where many Palestinians still consider all Israelis as eternal "enemies." Or in Iran and other Middle Eastern countries, where all Israelis are considered evil, where the "great Satan" is the United States.

Fortunately for the Middle East, some leaders are coming to their senses lately, and they find more importance in peace than in a continuance of religiously inspired hostilities and the status quo. The handshake between the Israeli prime minister and the PLO chairman on the White House lawn was one of the first signs of a shift in priorities, a shift from religious prejudice to common sense. However, a lot more logic has to be applied yet in the Middle East by persuading extremists to change their fundamentalist religious attitude.

Jewish settlers claim a "biblical" right to Hebron, by insisting that the prophet Abraham bought the land centuries ago for 400 shekel. Historically, Hebron served as the country's capital at one time and it is the burial place for Abraham, his wife, son Isaac and grandson Jacob. Compare that to the North American Indians, who ruled our part of the world and when the Mayans and Aztecs were the only inhabitants in South and Central America. Because of ancient heritage, do the natives of America demand all the land back that they called their own a thousand or more years ago? Of course not, peace and harmony must come long before biblical and traditional rights and before harm is inflicted on neighbors. And before revenge chain reaction is triggered once more.

Not only is there little difference in the physical appearance of the population of the Middle Eastern countries, many of their foods and customs are equally very similar. Even the differences between the Hebrew and the Arabic language are not all that great. Although they all believe in the same God, the main source of conflict originates from religion, or the interpretation of it. Moslem (Palestinians) believe THEIR religion is the only right interpretation of the written word. Jews believe only THEIR version is correct, and the Christian Palestinians insist, that THEIR belief is the only one to live by. All of them believe in their religious traditions, mainly because their parents and ancestors had that particular belief also. For example, it was and still is officially forbidden for a Muslim to marry a Jew or vice versa.

We could apply this observation again to India and to Lebanon, where even now in the name of ancient, unproven religious teachings, people kill each other usually ONLY because of differences in religious belief. Baby girls are killed after birth in areas of India only because of an ancient tradition, which requires women to bring a dowry to the future husband and his family. It is therefore quite acceptable for even the most religious natives to murder baby girls in India, when at the same time, the murder of a boy is a serious criminal offense and subject to severe penalties.

WHO WAS JESUS CHRIST?

Christians BELIEVE that Jesus Christ is the son of God. There seems to be no serious doubt by other major religions or by most non-believers about the existence of Jesus. However, many followers of other religions do not believe that Christ IS the son of God. Christians and non-Christians do understand that Christ, whoever he was, was not born around Christmas time and not in the year "one". According to some sources, it was around September and definitely in 7 B.C. It also happened during one of the rare years of a "triple conjunction" of Jupiter and Saturn when as seen from Earth, both planets line up perfectly behind each other, thus creating a very bright object in the night sky. When it happened three times during 7 B.C., it was thought to be a new star, the star of Bethlehem. Of course, it was not a "star" at all, and many baby boys were born during that year and the years before and after.

There is also no way to prove that Jesus was born to a VIRGIN mother. After all, Mary and Joseph WERE of opposite sexes and seen together often at the time, just as millions of young couples of opposite sexes still live together 2000 years later. Here, we must consider again the TIME and the PLACE in history, from where this Christian belief originates. It would be difficult to convince anybody TODAY, that a child was born as the "Son of God" to a VIRGIN mother, while in co-habitation with a male partner.

In the words of the great Thomas Paine 200 years ago: ". . . When also I am told that a woman called the Virgin Mary, said, or gave out that she was with child without any cohabitation with a man and that her betrothed husband, Joseph, said that an angel told him so, I have the right to believe them or not. Such a circumstance required a much stronger evidence than their bare word for it; but we have not even this — for neither Joseph nor Mary wrote any such matter themselves; it is only reported by others that THEY SAID SO — it is hearsay upon hearsay and I do not choose to rest my belief upon such evidence.

It is however, not difficult to account for the credit that was given the story of Jesus Christ being the Son of God. He was born at a time when heathen mythology had still some fashion and repute in the world, and that mythology had prepared the people for the belief of such a story. Almost all the extraordinary men that lived under the heathen mythology were reputed to be the sons of some of their Gods . . ." I could not have said it better myself, only in more modern English!

As mentioned, there is no doubt in the minds of most non-Christian believers, agnostics and even atheists that this person really existed. He was probably very talented and may have displayed, what was then considered exceptional powers. He may have preached to the people of Palestine all about good and evil during his short lifetime. Also, there is little doubt, that Jesus DID help the poor and that he convinced some of the rich traders of the day to sacrifice some of their wealth and distribute a portion amongst the poor. Islam teaches the same rules to the point that a follower of that faith should sacrifice about 10% of his wealth to the underprivileged.

Some Communist leaders of the seventies and eighties are known to have identified Jesus Christ as the first true Communist. Considering it all, they may not have been all that far from the truth with their statements. Or if someone compares Jesus with Robin Hood, it also would come close to reality. However, to be born to a VIRGIN mother who has a partner of the opposite sex, is an unacceptable story for the realist at the end of the twentieth century. Of course it is difficult after 2000 years to prove or disprove all, or some of what has been taught about Jesus.

But just as scientific theories are questioned, the true identity of Jesus should also be questioned, along with all other religious teachings. Jesus and his followers may have truly BELIEVED he was sacrificing his life for the better of mankind when he was about to be executed by crucifixion. Many so-called "martyrs" die today for a cause, often as a victim of strong BELIEF also.

A lot of research has been done on the identity of Christ, by theologians and others. Here are some of their conclusions:

Thomas Jefferson wrote a biography of Jesus, keeping most of the teachings but he discarded most Gospel passages, which he did not consider authentic.

The dutch Catholic theologian Edward Schillebeek stated that the New Testament is the testimony of BELIEVING PEOPLE only, after he completed his research.

The Protestant critic Rudolf Bultmann of the University of Marburg, Germany contended by 1926 that it was fruitless to further search for a "flesh-and-blood Nazarene beneath the myth" and "that we can now KNOW almost nothing about the life and personality of Jesus Christ."

The Presbyterian Reverend Burton Mack of the School of Theology at Claremont, California suggested that Jesus may have been executed by mistake. The Romans rounded up dissidents frequently and executed them by crucifixion without trial.

It has also been suggested that when Jesus said he was the "Son of God" he may not have wanted it to be taken literally. New Testament language of this type in referring to Jesus as the "Lamb" or "Word" of God is metaphorical (figurative).

Other historians suggested that Jesus had come from

the wilderness with the intention to preach the gentle message of peace and brotherhood to the frequently warring natives of the land. He may have been a revolutionary. The Gospels Matthew, Mark, Luke and John originated from second- and third-hand information which had been passed on over several decades, before it was written down. At that point, little could be proven as fact.

The lack of factual knowledge has led to the incredible number of Christian churches, confirming that Christianity not unlike most other major religions is based on mostly mythical hearsay. Whenever another scholar of Christianity transforms BELIEF to FACT, a new version of a Christian church or cult is created.

When it comes to the mythical resurrection; we have no facts, only the belief of a few devoted followers. Recently, even some Christian church leaders doubt that the resurrection really took place — that it was "factual."

Interesting enough, the contents of the New Testament are frequently questioned by theologians. One recent example is the statement from a German professor of theology, Dr. Gerd Luedemann, who interprets the earliest versions of the New Testament as contradicting the resurrection of Jesus. His research through the old scriptures ended with the conclusion that the body of Jesus actually decomposed after biological death in the same manner as that of any other living creature. Only visions of Jesus' followers introduced the myth of resurrection, according to his findings. Of course, his belief in Christ is unchanged, but his point is made subconsciously: one must question everything that does not make sense, including the various versions of the Bible.

Finally more quotes from Thomas Paine: "The Christian Mythologists tell us that Christ died for the sins of the world, and that he came on *purpose to die.* Would it not then have been the same if he had died of a fever or of small-pox, of old age, or of anything else?"

Elsewhere in *The Age of Reason* he has more to say about it: "But the Christian story of God the Father putting His son to death, or employing people to do it (for that is the plain language of the story) cannot be told by a parent to a child; and to tell him that it was done to make mankind happier and better is making the story still worse — AS IF MANKIND COULD BE IMPROVED BY THE EXAMPLE OF MURDER; and to tell him that all of this is a mystery is only making an excuse for the incredibility of it." I could probably not have said it better!

ALL LIFE IS REALLY "VICTIM" OR "BENEFICIARY" OF NATURAL OR MAN-MADE CIRCUMSTANCES

Based on scientific knowledge, all present life forms are located where they are as a result of NATURAL circumstances and evolution, as well as the driving force of life — Sex . . . As stated earlier, all life on Earth has managed to survive and evolve to it's present form as a result of the location and the events on the planet in relationship to our own star, the sun. I will go into that later. I am not a "Sun God" worshipper, but I can understand ancient traditional belief in "Sun Gods," many of today's tribal cultures in Africa and South America still worship the Sun. Some remote native tribes have never been infiltrated by Christian or Moslem "visitors" (usually — "Conquerors"!) and they are probably better off without "forced" major world religions. I will refer to tribal religions later.

Now I want to emphasize the importance of CIRCUMSTANCES, which are shaping our lives and of which we are ALL "victims" or "beneficiaries." I call it "the hit-and-miss lottery of life," or nature's "law of probabilities," along with natures' best known law, the law of "survival of the fittest." As stated earlier, all life on Earth is primarily controlled by the Sun's distance from Earth and by the amount of heat and other radiation reaching us from it, and its effect on microorganisms inside us and around us.

ALL living organisms depend on controlled quantities of solar radiation and are the "beneficiaries" or "victims" of these natural events or "circumstances." Good or bad, without them we would not exist, at least not in OUR solar system. Short-term and more or less predictable circumstances such as climatic changes, alterations in weather patterns or natural disasters, influence our life and that of our global biological fellow inhabitants almost daily.

Most changes in weather and climate are caused by the effects of changing solar radiation, we have overwhelming scientific proof of it. Our ancestors took the word from the Bible and believed that God was "punishing evil sinners" through severe drought or floods, without their understanding of the importance of the Sun. Sometimes, it appears that more ancient civilizations had better knowledge about the importance of our Sun than more recent "civilized" ancestors.

Long-term conditions, such as ice ages and continuous drought effect life forms in much greater numbers and for much longer time periods. Again, these events are at this time usually predictable through the various sciences and therefore avoidable by evacuation, in the worst case scenario. No evidence is apparent of God's "Punishment of sinners" after a natural disaster wipes out thousands. 2000 years ago, humanity was not able to foresee long-term major climatic changes to large parts of our planet. It was easy to explain to the masses that Allah was behind it all.

When the last ice age in North America started to retreat about 38,000 years ago, (about 12,000 years ago in Northern Europe), all life-forms in the effected areas became "victims of circumstances." Those who managed to leave the effected areas saved themselves and future generations. Those who stayed behind, either became extinct or had to be able to adapt to the somewhat radical changes, through evolution. In either case, all life forms became "victims of circumstances," including the plants and trees which were unable to "move" to a warmer climate. But even plant life was not all lost, since some of the seeds moved with the advancing glaciers and ended up thousands of miles further south when the glaciers retreated. Or the seeds remained frozen in the glaciers until the ice melted again thousands of years later. In summarizing it, the evidence is unquestionable: all life-forms are "victims" or "beneficiaries" of matter, time, space, gravity and other natural circumstances.

Earlier, a much longer lasting ice age affected Europe for 900,000 of the last one million years. A suspected wobble in the Earth's rotation may have caused most of Europe to be covered by glaciers. Was it God's "punishment" of evil sinners? Of course not, there were no civilized people in Europe at that time. It was only about sixty to thirty thousand years ago, that Neanderthal man inhabited that part of the world. "Modern" man appeared first about 25,000 years, also long before the "modern," three thousand year old religions were known.

However, let us look at some man-made and more individual "victims of circumstances": The unfortunate child, having been born three, four years ago in Bosnia to Muslim parents, who lost his father in battle and whose mother was raped by Serb "Christians" is a victim of such man-made circumstances. Strong religious belief was no help — it was the CAUSE of these tragic events. In fact the child had no say in the matter.

The survivor of a kidney transplant is more fortunate — he is a "beneficiary of circumstances." He or she would have died as late as the sixties, before transplants became routine. The patient is fortunate to live in a technologically advanced country, thanks to our more recent ancestors. Once again, no sign of help from a superior GOD. Only FACTS and modern medical science have played a life-or-death deciding role in the survival of this person.

Following is a typical and tragic case of a "victim of circumstances," a true story about a young man: A young machine operator had an inflamed, sore throat. His family doctor gave him some penicillin shots and the throat infection almost disappeared. However, about 3 days later, our young friend began to itch and started to develop hives all over his body. The doctor's immediate diagnosis was that the reaction was caused by an allergy to penicillin. He prescribed antihistamines and the hives and itching nearly disappeared. The young man experienced drowsiness from the antihistamines and, seemingly as a result of it, accidentally cut his hand at work. The plant nurse applied an antibacterial ointment which contained penicillin and the hives and itching returned. This time the young victim also showed swelling of his lips and eyes.

The doctor used an application of corticosteroid treatment in consideration of the potential dangerous effects of the more serious penicillin allergy. The allergic reaction nearly vanished again but the patient developed abdominal pains and heartburn, and showed signs of blood in his stool. Even though the diagnosis was made correctly, the introduction of the corticosteroid had induced a peptic ulcer.

Two physicians agreed to operate and to remove the ulcerated portion of the stomach, and in the end the operation was successful. Due to an unavoidable loss of blood during the operation, the patient had to have a transfusion of two pints of blood. As a result of that, he developed hepatitis and jaundice and had to be fed intravenously for a few days. He recovered quickly. However, from the intravenous injection of the plastic tube, he developed an inflammation around the area of the injection. Since the doctors now knew very well how allergic the patient was to penicillin, tetracycline was injected in place of it. The inflammation disappeared, but the patient now had severe cramps and diarrhea from the antibiotics.

A special diet and a new synthetic and antispasmodic drug was prescribed, and the diarrhea together with the cramps vanished. The new drug relaxed all muscles and caused dilation of the patient's pupils, with resulting impairment of his vision. He drove his car into a tree and was killed instantly. By the way, he was an average believer, not particularly religious, but he DID attend church regularly and was no more and no less a "sinner" than millions of other "victims of circumstances."

Getting back to transplants of body parts and to blood transfusions, religious scripts have taught us very little about the realities of the human "animal's" body or about transplants and medical science. In many instances, we have only been misled and misinformed about scientific facts through ancient religious teachings and morals. Blood transfusions and transplants of body parts would have been a deadly "sin" and interpreted as "meddling with God's will," or labelled as "witchcraft" only a few hundred years ago. Transplants are now of course acceptable and a common occurrence, only thanks to advances in medical science.

No instruction for transplants can be found in the scripture. Interestingly enough, we are now going as far as starting to use animal body parts for experimental transplants! Although it proves our close relationship to the rest of the mammal family, it is even less acceptable to most religions than transplants from one HUMAN to another. Therefore, survivors of organ transplants are all "beneficiaries" of circumstances, originating from rather unfortunate events, usually a fatal accident of another human being. The fortunate circumstances for the survivor are to be alive in a scientifically advanced country at the end of the twentieth century, and to be able to afford the cost of a transplant.

"VICTIMS" AND "BENEFICIARIES", REGARDLESS OF RELIGIOUS FAITH

During my various business related travels to wealthy AND to poor countries, I saw and spoke to both, UNFORTUNATE victims AND to rather FORTUNATE "beneficiaries" of circumstances. I saw begging children in South Asia, who had a foot cut off to arouse sympathy to passers-by. Surely a victim of circumstances, born into the only "hell," which I can believe in, because I can see it, it is FACT.

The children and adults of Ethiopia, Somalia, and the Sudan, as well as those born in many South American countries are victims of circumstances. Often "hellish" circumstances. Only a few hours away by plane, the child of a wealthy Arab businessman or the child of an upper middle class American family are also the result of circumstances. Much more fortunate ones.

We previously looked at WHY we are WHERE we are as a result of a series of natural and historical events, or "circumstances." If our parents were born in Europe and decided to emigrate to the USA or Canada, or our grandparents came over, that is WHY we are HERE, and not in Germany or in Italy, or in Mexico. If our parents were relatively well to do we had a good head start, but most important, we live in a land where opportunities exist. Not so in Somalia, Bangladesh, or a few years ago in Lebanon. Just to name three examples of unfortunate "circumstances," or locations.

However, the place of birth and apparent good fortune through inheritance, or business success, can also result in tragic "circumstances." In 1993 two popular baseball players were killed in a boating accident in Florida, becoming victims of tragic circumstances. Without fame and money, they may never have had the fatal boat ride and could have lived a long life, of modest means maybe, and possibly with families.

The news media reports daily about traffic accidents, industrial fatalities and gun-related injuries and deaths. All fatal victims are "victims of circumstances," to be exact in some cases: "Victims of gravity." However, had these victims been living in a developing South American country, they may not have owned a car or may not have worked at an industrial site. As a result of it, they would still be alive. Maybe in poverty, but alive.

If a Haitian "economic" refugee manages to leave the poverty of his homeland behind and when he starts a new life in North America, his past, present and future are a combination of man-made circumstances regardless of religion. It was not prayers which helped him, only his own determination and effort, together with "lucky circumstances" when he is successful. When he can afford to start a family in his new homeland, his children in turn now become much more fortunate "beneficiaries of circumstances." He "created" the better conditions for himself and his new family, regardless of his religious belief, if any. HE made the better "circumstances" on his own.

Had one of the annual estimated 30,000 North American fatal victims of crimes or accidents been living in Japan, where the possession of a gun is unwanted, most difficult to arrange and not "in," a potential gunshot victim would still be alive. These tragedies were obviously circumstantial and not arranged by fate through "God." One of the largest and most populated cities on Earth is Tokyo with a population of about 10,000,000 souls. About 10 people die annually through gun related injuries in Tokyo. Let me compare this to New York City with about the same population, where 1,672 people died a violent death in 1987, and by far most of them through gun related injuries.

The religious belief (if any) of the victims or even that of the killers had no influence whatsoever on this outcome. Lucky to be born in Tokyo, another 1,660 people are still alive there this year with similar numbers following every year. Plus their future generations, and so on. All victims of man-made circumstances, tragic ones in New York City and more fortunate ones in Tokyo.

The murder victim from a happy, close knit and religious family living in an American city may have been born a beneficiary of fortunate circumstances. But by being in the wrong place at the wrong time, he or she became a victim of a killing during a jewelry store robbery . These are undisputable facts. The victim's religious belief and even his or her character made no difference to the causes of the tragic end. Another victim of circumstances.

During a 1993 tragic automobile crash in the Canadian Province of Quebec, nineteen elderly people died a horrible, fiery death. Many of their bodies were unrecognizable and it was one of the worst crashes in Canadian history. Yet, something became most ironic for this man-made disaster — the unfortunate seniors were returning from a religious pilgrimage! Later, the victims' families were shown during a TV newscast — praying in Church! What powerful religious faith can do to the human mind! In reality, it was just another case of "victims of circumstances," or to be precise — victims of gravity, location and time.

After the last California earthquake, a young man was interviewed on national TV. His plight was such that it could be identified as the typical "born loser's" story. He lost a restaurant in L.A. during the riots, when it was severely damaged with the resulting loss of regular customers. He then established his business in Malibou, where it was totally destroyed during the recent fires, and finally, when he set up shop again in another location in Southern California, the building was a total loss again during the earthquake.

I picked this episode to illustrate again that it was NOT "God's" punishment of an evil person and that he was simply just another unfortunate victim of circumstances. During the TV interview, he certainly did not leave the impression of being an "evil" man and he was particularly happy, that he and his three-legged dog were still alive. He had not given up hope, and he realized, when one chooses to live in southern California, one takes a chance with earthquakes and fires. Just as residents of Florida take a chance on hurricanes and as the locals in the north eastern U.S. and in Canada know that there will be severe frost and snowstorms at one time or another during the winter.

When a physically inactive, older person with a weak heart in Michigan decides to tackle a six foot snowdrift with a snow shovel, he or she is taking a chance of collapsing with a heart attack. If the tragic end really comes at that time, the victim is strictly a "victim of circumstances," without evidence that a higher "God" chose the time. When tens of thousands of smokers die annually because of smoke-related diseases, they also are (self-made) victims of circumstances and just as in the previous case, their early death was preventable. It was their own risk taking which cost them their lives. No different than a race car driver taking a chance — No involvement of a "God" or for that matter, an equally mythical "Satan."

To illustrate blind religious faith again: During a TV interview, one out of five victims of the Mississippi floods in 1993 were convinced that "God wanted to punish sinners". At the same time the tidal waves from the most powerful Tsunami earthquake on the northern Japanese island caused about 100 deaths. Had it happened in Tokyo, the devastation and loss of human lives would have been astronomical. But since the Japanese Islanders are mostly non-Christians, probably not one of them would have made a statement such as the victims of the Mississippi floods. Considering they are not Christians, they were very lucky in getting away with only 100 deaths.

Circumstances control our lives in many ways. For example, chances are the victim of a sex assault was YOUNG and FEMALE and was unknowingly at the wrong place, at the wrong time. The attacker may have created unfortunate circumstances by being under the influence of alcohol or drugs, or he may have had an extraordinary sex drive due to a possible imbalance of X and Y chromosomes. It was a very tragic combination of circumstances and had nothing to do with religious belief of either the victim or the attacker. Other "victims of circumstances" are young boys and girls under the care of adults and teachers of religious institutions. Lately, we read and hear about cases of child molestation in religious schools more often than we care to.

SCIENTIFIC FACTS AND CHANGING ENVIRONMENT ARE SHAPING OUR LIVES, NOT RELIGIOUS MYTH

Victims of genetic diseases are a medical fact, and since we have no choice in the selection of our parents, we have to accept the good with the bad. As an example, I personally consider myself fortunate in this respect — both my parents were relatively healthy individuals and both died at a ripe old age. However, my father died as a result of a heart attack at age 74 in the early sixties. Had he known more about the effects of cholesterol and other potential causes of heart disease as much as we do today, AND had he taken the necessary precautions, he may have lived ten, fifteen years longer. Again, medical SCIENCE has taught us preventative measures, NOT religion.

We now know for fact that a balanced diet and certain supplementary vitamins and minerals will prevent diseases and will in all probability extend our life. More than we can say about prayers. And if prayers really "appear" to help, one could ask the question: was it the prayers that cured an illness or was the strength of belief and willpower to live, or could the illness have been eventually eliminated anyway by the body's own resistance?

Medical research also has recently given us the knowledge to avoid pottery, glass and crystal containers with a high lead content. Until humanity had that knowledge, thousands died early as a result of lead poisoning or liver ailments caused by excessive lead intake. Evidence has been found to that effect in the remains of ancient Roman skeletons. Their lives had been significantly shortened through the use of lead or tin-based water pipes and pewter containers, causing early and usually fatal liver disease. Along with other preventable causes, a number of diseases shortened the lives of our early ancestors, only due to lack of scientific knowledge. The common non-anodized aluminium cookware of past years has been a possible major contributor to such illnesses as Alzheimer's disease. Once again, religious scriptures have not warned us about it.

On the other hand, we HAVE the knowledge now on the causes and possible prevention of a number of life-shortening illnesses, but many of us choose to ignore them (high cholesterol junk food, smoking, drugs). Our lives are "created" by our parents, grandparents and our ancestors through the reality of conception and birth. We know this to be fact. The quality of life is determined the same way, not by a hypothetical superior being as daily events teach us reality. Recessions and unemployment are man-made circumstances which effect the life of a great number of people, religious believers and nonbelievers equally. With or without prayers.

A typical family history would be the following scenario, cause and effect of the "circumstances" to the lives of several generations: The European great grandparents of a North American man met about 1885 on an ocean liner. He was the captain, she was a chamber maid. Circumstances brought them together, since they both worked in the same place. Happens every day in modern days also. Before they had a chance to get married, a little boy was conceived and shortly after his birth, the father (captain) died. Not unusual in those days, many people died at a very early age because of a lack of medical health information, or, how fundamentalist believers would put it, "it was God's will." In reality, they were born 100 years too early.

As a result of the captain's early demise, no other children came out of this relationship. The chamber maid left the ship and lived with her family near the border between France and Germany. The little boy grew up and showed particular interest in mechanical things. Maybe he had this "talent" "inherited" from his father, the captain? He took an apprenticeship in plumbing and steam heating, and shortly after graduation, moved to the big city in Germany where opportunities for work looked more promising.

Before all this happened in the 1870s, another young

couple met in a nearby town, got married and had their only child in 1875, a baby girl. Unfortunately, the father of the little girl had a serious accident as a train engineer in his mid-fifties. He was badly burned and survived only a few years until his death after the turn of the century. A victim of circumstances of the times, since train locomotives were powered by STEAM. However, the only daughter had grown into a young lady by 1895, when she met and shortly after married a young man with a pretty secure job at the local post office. Two years later, their first child, a daughter, was born. Only three years after that (1900), a boy was born into this marriage.

Getting back to the captain's and chamber maid's son, he had left the plumbing business and was a salesman for a manufacturer of chandeliers and lamps in the big city. By 1929, he happened to travel in the same train as the postmaster's daughter. Again, attraction between members of opposite sexes set in and the relationship blossomed into marriage, a boy was born in 1931.

Only weeks before this boy was born, the postmaster died at the relatively early age of 61 of a massive heart attack after other minor illnesses. He had liked, as so many others of his time period in European history, greasy home fries, butter, sandwiches covered with lard and salt and rich dairy products. Little did anybody know at that time about the effects of high cholesterol and high calorie foods. This information could not be found in medical books in the thirties and certainly not in the Bible.

His only son completed university and accepted an engineering position with a German aircraft manufacturer. But he also became a victim of circumstances at the young age of 37, when he died as the results of a devastating, incurable disease of the time — tuberculosis. It was long before the discovery of penicillin. And of course, without surviving children and with the only male of the family now gone, the postmaster's name also died.

Long before divorces were a common occurrence in Europe or North America, the chandelier salesman and the postmaster's daughter had a divorce after only six years of a not-so-happy marriage. Therefore, the little boy from this marriage remained the only child out of this relationship. The mother had developed a goiter and after an emergency operation, had to move to another city away from mountains and closer to sea level. The little boy, his mother and the grandmother became victims of events again, because the operation and the move would not have been necessary if iodine would have been more freely available as it is now, simply in salt as a supplement.

Soon after that World War II broke out and the trio found themselves in a strange and large city in central Germany, on the "wrong side" of the warring parties, because they were all born in Germany, as were their ancestors before that, in a way again all victims of circumstances.

The young boy, his mother and grandmother lived through a lot of inconveniences first and severe hardships later along with hundreds of millions of other Europeans during and after the war. Friends and family members were killed. The numbers vary, but an incredible 45,000,000 people are said to have lost their lives during WW II, all victims of tragic, man-made circumstances. Forty five million families partly or completely destroyed and many victims never had a chance to start a new family before their untimely death.

Religious fundamentalists of all religions, ordinary believers, non-believers, good people and evil people died. Some of them under horrific pains — yet others died so fast they never knew what hit them. However, there is no scientific or logical evidence that an almighty, mythical "God" or a "Satan" were involved in the fate of the millions of victims.

Churches and hospitals were destroyed just as regularly as armament factories, no Satan and no God was in sight during any of it. Fundamentalists may claim, that Hitler WAS Satan in person. My question to them is: if so, why did the "almighty" God allow him to be born in the first place? If the answer is: "God works in mysterious ways," I would reply: "To allow the birth of one "satanic" person, 45 million had to lose their lives?" To call the 45 million, "victims of (man-made) circumstances," makes more sense to me. The evidence is too overwhelming.

Getting back to our young boy in central Germany, just at the end of the war his grandmother died to cancer of the uterus at age 69. Again, modern medical tests and early diagnosis could probably have prevented it in later years. The young fellow decided to leave a rather hopeless situation, in what later became East Germany, looking for a bit more hopeful future in the West German city where his remarried father still lived. Unfortunately, the circumstances prevented him from getting the education he had wanted to have so much, as a biologist or an astronomer. He had to look for work in the food industry instead.

To cut the story short, he met a local girl in the late forties and an early marriage brought a total of five children. Two of them were born in Germany, the others in North America. He had become frustrated with the slow progress of the economy in West Germany during the early fifties and he chose Canada as the country for a better future for his family. Neither God nor Satan influenced him. The choice was between Australia (too far, not enough money for the journey), Venezuela (only Spanish speaking, he only had knowledge of high school English), United States (very difficult to get a Visa), or Canada, as the relatively easiest choice.

Now, as a result of the decision made by the young man and his early leadership to the rest of the family, his children and grandchildren are Canadians and future generations will be the same, unless one of them also decides to move to the United States or to another country, thereby changing the whole family history again.

Just one of millions of European stories of this century, with mainly "victims" and a few "beneficiaries" of the prevailing circumstances.

The moral of the story is once again: Human beings are the "result" of their ancestors' actions and genetic features and flaws, their own decisions, circumstances around them and the advancement of the various sciences. I see no involvement of a hypothetical "God," or for that matter, of "Satan."

The stories of "victims of circumstances" could fill endless volumes of books, because we are ALL victims of a combination of circumstances, good and bad. I hope, the previous few examples will illustrate to the reader HOW MUCH our lives are influenced by FACT and by CIRCUMSTANCES and NOT by a mythical "God", unless we become victims of a "holy war" or another religious conflict.

BIRTH, LIFE AND DEATH — DETERMINED BY "GOD" OR BY THE FACTS OF NATURE?

So far, we have failed to find scientific evidence which could back up RELIGIOUS teachings of how we were "created", how long we live and when and how we die. Whether we are suffering in misery or we are enjoying a good life, it is up to us, to the environment, to inherited good or bad health. It is all determined by scientific fact and by the predictable or by yet unknown laws of nature.

Also, natural disasters do not distinguish between religious believers and non-religious folks. No choice is made between "evil sinners" or religious fundamentalists. Floods and earthquakes kill Christians, Buddhists, Muslims, Hindus and all other believers, as well as non-believers. Victims are known to have died during an earthquake while praying right inside a church, while others (non-believers) survived outside. Sometimes it is the other way around. Of course, in the last instance we are told by believers: "It was a miracle!" While the first scenario would predictably be explained: "God wanted to bring them home to heaven," or words to that affect. All religions have an answer for everything, except the answers are seldom, if ever, backed up by scientific evidence or by logic.

All adults are familiar with the facts of conception and birth, originating from instinctive sexual animal attraction between our parents. Had records been kept, we could trace our ancestry back to the Roman empire, to the ancient Egyptians or Greeks. Archaeologists have found remains of advanced civilizations in China and in North America, more than 10,000 years old.

After many years of research, some geologists have now come to the conclusion that the Sphinx in Egypt is also closer to 10,000 years old. Previously, it was thought to be only about 4,500 years old. However, not all scientists agree on a 10,000 year old Sphinx. No explanation has been found yet how the builders of this gigantic structure managed to move 200 ton rocks to build the base for this huge statue, whether it was 4500 or 10,000 years ago. The point is: our traceable ancestors were around for AT LEAST 10,000 years.

Archaeologists have discovered skeletons of our "ancestors" in Africa dating back five million years, and primitive personal effects as old as two to three million years. However, no evidence or remains of the biblical Adam and Eve have been found. And if there was such a couple, first of all: were they Africans, or were they of Oriental, Asians or of white Mediterranean, Caucasian racial origin? Or were they Mayan Indians? Or maybe they were blond and blue-eyed Caucasians?

The existence of all the known human races long before biblical days turns the hypothetical existence of Adam and Eve as the first human beings into nothing more than a totally unfounded myth. Again, let us not forget, WHEN and WHERE the Bible was written. Only the races existing in THAT area of the world were known to the locals.

It would be just as easy to BELIEVE that our primary ancestors were shipwrecked aliens (for those, who believe in the existence of extraterrestrial life at all), as it is to believe in the biblical story about the "first people" on Earth. At least, we have an appreciable number of creditable eyewitness reports about alien spaceships (UFOs), even though we have generally no official scientific recognition as yet. But we have no scientific evidence, and of course no eyewitness reports about the BIBLICAL first inhabitants of this planet.

If we are not satisfied with the scientific evidence and still rather believe in the biblical explanation of the origin of humanity from Adam and Eve, one is left with another question, (in addition to the differences in appearance of the races): "Where did THEY come from?" If we consider alien visitations to Earth to be a fact, it would be possible that some of the first human beings on Earth WERE stranded aliens, which later interbred with the more primitive earthlings of the time. It may be a quite conceivable theory that some of the stranded or visiting aliens came from various different solar systems and, after combining with local primitive humanoids, created the various human races. A theory, that makes a lot more sense than "Adam and Eve."

NEGATIVE AND POSITIVE EFFECTS THROUGH CIVILIZATION AND TECHNOLOGY

Humanity has polluted the waters, the surface and the air of our planet faster and more effectively in the past 100 years than during the 10,000 years prior to that. That is no secret. Fortunately, we have also discovered ways and means to prevent further pollution or even to reverse it. Almost all pollution in recent years was and is caused by man-made chemicals and imbalances such as massive deforestation.

On the other hand, and as stated earlier, scientist have discovered bacteria colonies which thrive on certain pollutants. Such bacteria will eventually "digest" the largest oil spills, after some more research work is completed.

Research can now be applied to peaceful goals, after the threat of a major war had been eliminated with the collapse of the iron curtain.

Unfortunately, some "negative" events are happening too fast for the evolutionary process to adapt life forms to rapid changes, and it all came about without warnings from religious scripts. At the time when the Bible and the Koran were written, pollution was an unknown factor and it is therefore not mentioned in these scriptures. As a result of it, avoidance, elimination and reversing pollution had been left to modern science and ingenuity.

On the positive side, it is enlightening to see occasional small examples of FAVORABLE man-made changes to the environment. Take this small example: The government and inhabitants of the little known small, but oil-rich Arab Nation, the United Arab Emirates, have been converting thousands of acres of desert land into park and farmland. Through complex irrigation systems, seemingly useless desert wasteland has been turned into fertile agricultural land in a matter of a few short years. A number of other communities followed a similar path in Europe, North America and the Middle East.

Unfortunately, other nations in a similar situation do not have the funds for a project of this magnitude. Also victims of circumstances — along with all life in those affected areas, mainly in underdeveloped nations. An example is the destruction of ancient rain forests in South America, Malaysia and other developing countries, at an alarming rate of speed. Yet we forget, a lot of this destruction in done FOR the industrialized nations to supply them with lumber and other natural resources. Some of the exotic lumber may even end up as construction material for a new church, a mosque or any other new house of worship.

The end of the cold war was a very FORTUNATE man-made condition, after it had been created by humanity in the first place for the political dominance of their respective territories. Sounds a lot like other animal herds fighting over territory, except they do not have nuclear weapons to aid their cause. So, where was the so called human superiority over other creatures at THAT time? Or if there is any, was it not used for the wrong cause?

There are no biblical predictions about the end of communism in Eastern Europe. Instead, we can read about the coming of Armageddon, and of other "prophecies" of doom and gloom. Many religious groups thought they knew the exact date for this catastrophe, but their interpretation of prophecies never materialized. Some were openly disappointed, when civilization did NOT come to an end in accordance with religious prophecy!

DEATH — PART OF A NATURAL CHAIN OF EVENTS OR "GOD'S" WISH? WHAT IS "HEAVEN" AND "HELL?"

We looked at it before; HOW we got here through a long chain of ancestors, (unless some of us would still prefer to BELIEVE over scientific REALITY). Let us look at HOW and WHY we will all sooner or later DIE, believer or atheist, homo sapiens or other living creature.

Most religions teach us that it is God's will when and how we die, and if we will go to "heaven" or to "hell." Other than a number of eyewitness reports of ghost (or spirit) sightings, we have no scientific facts to prove or to disprove it. We have no scientific evidence to confirm the existence of "angels" and of a mythical "heaven." No one can prove to this day that our soul will leave the body at the time of our death, or that it will go anywhere other than into a grave with the body, underground, under water or into ashes.

It IS known however, some "invisible" information, such as so-called "instinct," IS genetically transferred to newborns from parents and recent ancestors. Certain habits in movement and posture, as well as good and bad physical features are known to be transmitted from ancestors.

Is it not conceivable therefore that the "spirit" is transferred in the same way as genes are from the ancestors? The evidence seems to point in that direction.

Medical science has developed a "map" of over 100,000 genes within each human body cell, the purpose of most of them is still unknown to us. Could it be, that just as our physical features, our "soul" is also genetically transferred? Until proof comes along as a better explanation, I tend to believe in this theory rather than believing in the biblical transfer of the spirit to a "heaven up there" or a "hell down there."

Other than physical features, we can observe frequent similarities between parents or prior generations and children, such as artistic talents, a like or dislike for music and other forms of art. Also inherited appears to be intelligence levels, self esteem, apathy, enthusiasm (or lack of it), optimism or pessimism, certain moods and frequent state of depression, schizophrenia, sadistic behavior (example: child abuse) and masochism, etc., etc. Just as physical and genetic features and faults are inherited, could spiritual faults and features not also be inherited?

We have more evidence to back up this possibility than we have of reincarnation or other mythical explanations about the origin and destination of our "soul" or "spirit." Of course, the acceptance of this idea would preclude the biblical explanation of origin and destiny of our "Spirits" — it would also eliminate the questionable myth of our spirits living beyond the grave.

One fact of nature is: nothing exists "forever," not even our solar system, — everything goes through a cycle of birth, life and death — Why do we believe our "soul" to be different? We are only left with one KNOWN FACT about one's spirit or soul: it DOES live on "Forever" in the memory of the survivors. The "good" soul of Mother Theresa will be remembered long after her death, just as the "evil" spirits of Hitler and Stalin are being remembered as such by millions, and for a long time to come.

The biblical imaginary fiery hell "down there" and

traditional artists' concept of a hypothetical "horned," red-skinned human-like image of "Satan," — half human, half horse — is certainly no proven fact. I referred earlier to a REAL living hell on Earth in some of the very underprivileged countries and to "satanic" behavior of certain individuals. But the suffering inhabitants were not born into "hell" because they were "evil." As we know, they are there because of circumstances. Of course, there are plenty of "evil" people on this Earth, if one wants to consider some of them as "possessed by (an invisible) Satan." In reality, they are also "victims of circumstances," often through inherited "evil" genes, mutations or because of the company they are keeping, or a combination of all.

Many accident victims claim to have had so called "near-death" experiences, ending up temporary in some sort of "hell." During a TV interview, a Sunday school teacher described such a terrible "experience" while clinically dead. However, nobody saw "Satan" himself, between a group of people, who all claimed to have had a similar traumatic experience. As it turned out during the conversation, all of them had experienced some sort of "low point" in their lives just before the clinical death. My logical conclusion was: their fear of having to go to hell triggered their minds accordingly.

The opposite seems to apply to persons who also, (during periods of clinical death) experienced the much more often reported relaxed and happy feeling of floating towards a bright light at the end of a tunnel and to "heaven." I agree with scientists who developed a theory about these experiences. Since the individuals were all only "clinically" dead and not "biologically", their brain still functioned to a certain degree and the memory was triggered about long forgotten and learned impressions about heaven or hell. Apparently, the victims usually were followers of either the Judea/Christian or Islamic faith if they "saw" the likes of a "heaven" or a "hell" during clinical death.

Anyone who had grown up in an area of the world

with a religious faith, other than the predominant one (or none at all), had either no memory or recalled some other kind of experience. Although an atheist reported to have turned Christian after "experiencing hell," he DID live in North America where he was subjected to local religions and where he was frowned upon by believers. Maybe because he WAS an atheist, did he have a build-up of fear about "hell" in his mind? To me, these experiences are dreams based on fear or on a fearful anticipation of an after-life.

Incidentally, the reports were suspiciously similar to stories told by drug addicts while they experienced hallucinations during overdose trips. It is known, that some people were so fanatic about their "near-heaven" hallucinations, whether it was drug induced or during a near-death experience, they actually committed suicide. At least they died happy in anticipation of going to heaven, their brain had "generated" the existence of a heaven.

The idea of a "heaven up there" and "hell down there" is simply unproven. Nobody has come back yet to tell us about life after biological death, and to prove how it really is to be in heaven or hell after death. To date, we have no proof of the existence of a "heavenly" planet in another star system, although it is a remote, unproven possibility. In reality, any other planet along with its own star would be subject to the same laws of natural evolution, environment and climatic condition as our own system. The conditions can never be "perfect," such as heaven is described to us in scriptures.

The known laws of nature do not allow 100% perfect conditions on ANY planet. Therefore, a "heavenly" place (planet?) could not possibly exist. If our souls would end up on this hypothetical planet in form of human beings, we would still be subject to the "good" and "bad" natural laws of birth and death, hunger, pain, sexual urges, good and bad feelings and everything that comes with the needs of a living creature. It will in all probability take astronomers many more years of intensive search before we find another "imperfect" planet similar to our own within the adjacent cosmos, capable of sustaining living conditions for intelligent, humanoid creatures. It is extremely logical and highly probable that only the vast distances in the Universe have prevented us so far from finding a matching or at least, a similar planet.

Expecting to find a "heavenly" place would simply contradict all basic and known laws of the universe and of nature. The existence of "Heaven," a perfect place for all those millions of people from this Earth, who according to religious belief deserve to live there forever in peace and harmony is pure fairy tale fantasy and wishful thinking. It may be something to look forward to, and to dream about. But I don't count on it, scientific knowledge tells otherwise. When it's over, it's over, unless my soul is "transferred" to another, new human being, and with a little luck not in a poor, developing country.

It must be emphasized again: the scientifically proven laws of nature do not allow "eternity" for anything, including a "soul." NOTHING lasts forever in its present form, EVERYTHING is subject to predetermined change, to the laws of physics and of chemistry. That includes the stars, the planets, all life forms, and the entire universe itself.

What we DO KNOW is how humans develop from an egg cell and a sperm cell like any other animal species, and how both cells carry all genetic information forward. From the day of conception, all animal bodies are controlled by microscopic chemical and biological "particles," or "micro organisms," and by a whole chain of natural and man-made events. All animal life, including Homo Sapiens, "inherits" everything through DNA and the genes from their ancestors, including health, appearance and character, along with many other factors.

We have to take the good with the bad again. Incurable diseases, such as Lou Gehrig's disease are genetically inherited, as well as GOOD health. Parkinson's disease can be traced through family history, and possibly triggered in recent decades by excessive intake of pesticides through the food chain. We don't have to go as far as genetic diseases resulting in early death: a simple example of hereditary "malfunction" is nearsightedness. Since one of my grandfathers and his son, (my uncle), were nearsighted, I had a good predetermined chance to inherit this common eye "malfunction" also. I did, and so did most of my children. On the positive side, none of my immediate ancestors were overweight through hereditary glandular malfunction. Therefore, I am also not overweight, regardless of how much I eat (Or ATE, when I was young). All my now adult children also do not have a tendency to be overweight.

Latest research has disclosed that breast cancer and Huntington's disease are in all probability also inherited through an ancestor's genes. Unfortunately, although medical science has this knowledge, we are not yet able to do something about it.

My question is to believers: If "God" creates us, why does heredity play such a significant role? After all, a parent with the genetically inherited Lou Gehrig's disease may be very "God-fearing," yet his unborn child has a very good chance to die at an early age from this terrible disease. Religion and prayers do not help, genes are more powerful.

"Genetic engineering" may be the only logical answer and eventually lead to a cure for most genetically inherited deficiencies. This branch of medical science has only sprung up recently and is often criticized by religious fundamentalists and believers. Unfortunately, they tend to overlook the fact that eventually millions of lives can be improved and prolonged through genetic engineering. They call the experiments: "playing God." I rather compare the beneficial evidence from genetic engineering with the "evidence" we have about the existence of a God, and how our lives have improved (?) through religious faith.

Recently, scientists have been able to probe deeply into the wonders of genetic engineering. We are now able to mix human genes with those of selected animals. Old Greek mythology left us with statues of "gods" of half human and half animal appearance. Maybe they knew something about the possibilities of genetic engineering THEN. Again, religious believers will reject genetic engineering as: "Man is playing God."

Looking at it logically, if genetic engineering helps to save lives or to improve them, if it does not harm anybody, if it makes just plain common sense — WHY NOT DO IT? Is it not more logical to worry about life and health first, before choosing man-made so called (religious) morals?

Medical science very recently discovered one more genetic factor, also inherited from our forefathers. This gene, "fragile X" is carried by about one out of 700 North Americans, and causes children to be born as "slow learners." This symptom can be mild or it can go right up to full mental retardation. Many young couples have a difficult decision to make — to either go without children of their own, or to live by the biblical rules forbidding birth control and abortion — when their child has a 50% chance of being mentally retarded at birth.

The recent Nobel Prize winner, Kerry Mullace was rewarded for his years of research with "gene therapy." His revolutionary discoveries deserve the greatest attention since they may change the lives of millions for the better in the near future. His work makes it possible to identify defective genes in a human embryo and eventually ALTER the DNA to eliminate such defects, long before the baby is born. Almost all diseases, except infections have a hereditary background carried by all of the 100,000 human genes. Recently, colon cancer and cystic fibrosis have been included in the list of genetic "defects" which can be detected at an early stage of development of a human embryo. At this point we at least have an early warning system available to us — sooner or later we may be able to make the necessary "corrections" through genetic engineering.

Medicine of the future may very likely be "preventative" genetic engineering or therapy, rather than through drugs as "cure." All indication go into this direction. Recently as an example, Canadian patients have been successfully treated for certain liver diseases through genetic therapy. However, since the possibilities for genetic engineering are endless, many religious believers already voiced strong opposition to it since it is "not ethical to predict the future," in their opinion. What IS their priority: saving lives or worrying about possible ethical conflicts based only on hypothetical religious belief and NOT on common sense?

The LOGICAL choice should be clear: Either knowingly giving birth to a carrier of defective genes, and at the same time taking a chance on unnecessary hardships for the child and the parents, or by applying logic through genetic engineering and therapy? There really is only one choice — logic. Again, we have evidence of genes influencing human lives, but no such physical evidence exists whereby a mythical "creator" causes or prevents genetic defects. When the Judea-Christian-Islamic scriptures were written, nothing was known about genes, DNA, mutations and all the other FACTS of life. It is fortunate for our generation that we have many factual answers before us with more discoveries being made at all times.

In addition to inherited genes, our medical chances for survival are based on the strength of our heart, the sensitivity of the skin and other body parts and organs as well as by the "inherited" environment. As a typical example, medical examinations for life insurance coverage always include the question: "What was the cause of death of your parents?" if they are no longer alive. Insurance companies use these and other statistical facts ONLY, NOT speculation based on biblical predictions. Nor do they distinguish for the purpose of setting premium rates between a "good" or "evil" person. With exceptions of course when a death row inmate requests a life insurance policy, his request will surely be denied, no matter how healthy he is or how healthy his family was!

A person with exceptionally healthy parents and ancestors has a better chance to live a long, healthy life, regardless if he or she have a "good" or "bad" character. The length of our lives is not determined by our belief or disbelief in God or Allah. Our inherited genes, the environment and our choice of food intake are some of the deciding factors of reality, together with good or bad living habits.

In addition to "good" and "bad" genes, too much or too little and the wrong kind of food, smoking, drug or alcohol abuse determine greatly "WHEN" and "HOW" we die, not the religion we believe in. That is, unless one lives in an area where a religious (or territorial) war is fought and ends up killed or injured as a result of it.

In 1987 proven and statistical chances how and when to die in North America were as follows, compared to statistics of 1900:

CAUSE OF DEATH /	**1987** % of total	**1900** % of total
All cardiogenic diseases (above incl. arteries, etc.)	38.1	12.9
All cancer related diseases	22.4	3.8
Diabetes	1.8	0.6
Influenza, pneumonia	3.2	10.4
Liver disease, cirrhosis	1.2	1.3
All other diseases	19.3	65.9
Total NATURAL causes —	93.2	94.9
Accidents, homicides, suicides	6.8	5.1

It is remarkable to note the shift in some CAUSES of death. The numbers of course, still add up to 100%. However, medical research has been able to increase life EXPECTANCY at birth since 1900 by about 26 years. Religious belief had no influence on these numbers and ancient scriptures were no help in increasing the life expectancy.

Other scientists argue that ACTUAL life expectancy

has only increased by five years over the past 90 years for anyone of age 65, taking in account the reduction of deaths through childhood diseases. At the same time, this apparently small increase in life expectancy can really only be attributed to advances in medical research over the past 40 to 50 years. Not bad, even if this rate can be kept up, and at one year increase for every decade, our descendants could reach a ripe old age of 180 years by the end of the twenty-first century.

In reality, at the present speed of medical research, one can be much more optimistic. One of the major mysteries to be solved is the reduction of DHEA Hormone production at advancing age, this condition appears to be the main cause of age.

No matter how we look at it, the fact is that only 41 out of 100 people lived to be 65 in 1900 and by 1990 almost twice as many, 79 out of 100, to be exact. And not thanks to prayers or religious belief. Science has eliminated or controlled many diseases and made us aware of preventative measures. Unfortunately, with the newly-found knowledge at hand, a great number of early deaths are still self-inflicted through overweight from the wrong diet, smoking, little or no physical activity, drug and alcohol abuse and avoidable accidents. Also, protective vaccination against a major disease has been ignored or is not available at times, resulting in the unnecessary early demise of thousands of years.

Every death can be explained through scientific facts, and fatal accidents can be backed up with statistical numbers, all based on risk factors or "hit-and-miss," such as being in the wrong place — gravity or velocity — taking chances — climate — inappropriate use of firearms — violent crime — lack of fire prevention and the list goes on.

Following are examples of deaths per one billion people during a one-hour exposure to the following risks:

Living in an area where rabies is prevalent	0.4
Being vaccinated or inoculated	1.3
Bus or rail travel	80.0
Child asleep in a crib	175.0
Automobile travel	1,040.0
Boating in a small boat	1,325.0
Air travel	2,400.0
Cigarette smoking	2,600.0
Mountain climbing	2,700.0
Swimming	3,500.0
Motor-cycle racing in the UK	35,000.0
Alpine rock climbing	40,000.0
Professional boxing in the UK	70,000.0
Being born	80,000.0

Other statistics are: being killed by a shark, while at a beach in shark territory 300 million to one or just being attacked by one, 100 million to one, compared to being audited by the IRS at 66 to one!

Getting back to our first statistic, the remarkable increases in heart disease as a cause of death since 1900 is to the greatest part, the North American population's own fault. The greater availability of high cholesterol and processed foods and the decrease in physical exercise at the same time is obviously in direct relationship to the dramatic increase in cardiogenic causes of death since 1900. The same can be said about cancer, where we may add increasing man-made chemical pollution as a major contributing factor. Also, people dying of heart failure in their eighties today would have not lived that long in the first place eighty years ago.

The enormous decrease in cause of death through "other diseases" is again thanks to medical research and vaccinations against a number of formerly incurable and mostly infectious diseases. Surprisingly, death rates through man-made violence have only increased by a mere 1.7% over a period of about 87 years. Again, we have the evidence in front of us that we are victims of circumstances from the time before birth, all through life, right until death.

Next we have "accidental" death as a result of "circumstances." Example: A child victim of a drowning exposed himself more to the danger of drowning by being too close to the deep end of a pool, being a non-swimmer and not wearing a life jacket, being unsupervised or alone, living in a house with a pool or visiting a neighbor's home with a pool, etc., etc. When we remove any or all of these conditions, the child would either be still alive or would have had a much better chance to survive. Truly a victim of circumstances, just as the person seeking refuge under a tree during a severe thunderstorm. The examples of how accidental deaths could have been prevented are too numerous to count.

Other than man-made events and natural disaster, micro organisms strongly influence our bodies from conception until long after death. Our own antibodies and "friendly" bacteria fight a continuing battle against intruding, harmful micro organisms. Most of the time, and until we die, they win every battle. Without "good" bacteria, all animal life would be unable to digest food. Until we are able to discover the micro organisms, which can successfully defeat deadly viruses such as HIV and how to stop the growth of cancer cells, the fight against premature death through these two causes will be lost. Yet only fifty, sixty years ago, a much greater number of lives were regularly lost to tuberculosis and polio.

Since the discovery of the polio vaccine by Dr. Jonas Salk 40 years ago, maybe millions of lives have been spared by it. Here is another individual, who in my opinion could be considered a "Saint" in the true sense of the word. No religious Saint could possibly ever have saved as many lives as he did.

Just recently a report was published by British researchers that a molecule causing rheumatoid arthritis has been successfully isolated. Several test cases had overcome this crippling disease in a matter of weeks. Much additional testing has to be done, but the results are encouraging for approximately one percent of the European and North American population who had been suffering up to now in agony. Again, it was a simple molecule (microorganism) causing this widespread disease. It effected religious believers and non-believers, "sinners" and "saints" equally, with no sign of a "God's" will to "punish" or to "test" his "subjects."

Once again, medical science found a possible cure without help from a scripture, the odds of suffering and dying when and how are the same with or without religious belief, with or without prayers.

SCIENCE MAY NEVER BE PERFECT, IT IS USUALLY MORE BENEFICIAL TO HUMANITY THAN RELIGIOUS FAITH

Occasionally and unfortunately, scientists fail (medical in the following case). Trying to constantly stay ahead, errors are to be expected and some trial and error experiments end in failure. One case is particularly tragic when over 12,000 hemophiliacs in North America were infected with HIV virus infected blood clotting "factor". Unfortunately, not enough attention was paid at the time of transfusion — it may have taken HIV virus contaminated blood from ONE donor only to infect all 12,000 victims. There is no margin of error when applying the latest medical technology. Although, there is no excuse for it, religious indoctrination on the other hand directly and indirectly has caused a much larger number of deaths over the past 100 years alone.

Even today, believers are still influenced by various religions, often putting medical doctors into the awkward position of having to decide between life and death, or between religious "ethics" and medical logic. Case in point: A Dutch physician is frequently treating child patients with severe cases of Down's Syndrome. He does not extend the lives of absolute hopeless cases. His opinion is that a small child victim will not survive very long anyway and during a very short life will be unable to do the simplest tasks. He will however assist a child born without extremities since the mind is perfectly normal and the rest of the body is healthy. The missing arms and legs can be substituted with artificial limbs to allow a near normal life for the child. All decisions based on simple logic.

Naturally, his decisions cause a lot of controversy, particularly amongst religious believers; they claim he is playing "God." Others insist it is his duty to assist until the patient actually dies. However, abortion of a severely retarded or deformed fetus may be a preferable solution, as long as he has a written agreement from the parents.

Although we have a lot to learn and discover at this time, modern medicine has found many antidotes to assist our bodies and their defense mechanism. As a result of it, a significant number of deadly viruses and bacteria were successfully eliminated, or on the verge of it, such as polio and even certain types of cancers; TB and yellow fever are other well known examples. For those interested in numbers: the polio virus' dimension is 1/25 billionth of a meter.

Medical science has recently discovered microscopic "helpers" in form of antibodies and friendly bacteria to allow a healthier and longer life. We have no evidence of help from a "superior" creator or from scriptures. We do know however, penicillin, hormone injections and transplants are only three examples of how millions of lives have been saved, or at least prolonged for many years. We cannot find instructions in the Bible or in the Koran how to cure any disease, instead we are told "it is God's will." Only medical research teams had to in the past, and still are continuing to find the cures.

If a convicted rapist and murderer has healthy genes, he could live a ripe old age in prison. If a "God fearing" religious person with an outstanding character has healthy genes from his or her ancestors, he or she will ALSO live a healthy and, in this case a productive life. With poor genes from their parents, the same two individuals would expect to have short, lives in ill health and spend a lot of time in doctor's offices and hospitals and in possible agony. Regardless of character or religious faith.

One can read or hear about a new antidote or vaccine against another major disease in the news media on a regular basis. Millions of lives have been saved by medical research. Millions also live longer lives and have children, which would never have been born had mankind left life and death to religious teachings and to "God." It was properly applied human intelligence that saved lives.

A recent announcement in the news media referred to a vaccine which had been successfully developed to conquer malaria. Here we have a story about a real "saint," but without religious overtones: A Colombian researcher had started a number of years ago developing a vaccine against Malaria, the most devastating killer in the tropics. Up to 3.5 million natives of mainly tropical countries die annually as a result of the dreaded disease. Of course, they die regardless of their religious faith. Up to now, religion has not been able to save any of these millions. Yet, medical science had been equally as unsuccessful in succumbing malaria.

Not too much attention had been paid to the Colombian researcher in the industrially developed world since the majority of victims are poor natives of underdeveloped countries. However, he may possibly change all that, since he now has the support of the Colombian Government and North American medical professionals are recognizing the results of his work. The vaccine has been successful on laboratory animals and lately also in human test cases. His vaccine could soon save millions of mainly young lives, yet very little has been said about it in the media.

What makes him a living "saint" in my eyes is not only the relentless work in perfecting the vaccine, it is even more so his rejection of a 68 million dollar payment offer for the patent by a US drug manufacturer. He is literally giving up that amount of money with the explanation, that his vaccine would cost about ten times as much if he accepts the payment and it would therefore not be affordable by patients in third world countries. As a matter of fact, administering the vaccine to millions of potential victims in Africa, Asia and South America would have to be on a "no-charge"-basis in most cases. It is easy to compare that with the imagined "success" of so called faith healers on other, much less harmful diseases but with a lot more "propaganda" coverage in the media. By the way, have faith healers ever cured a person dying from malaria?

Recently, a Chinese herbal drug also has been successfully used to combat malaria, and it too appears not to have immediate side effects. That is the good news, the bad news came when so called "super bugs" had been discovered, viruses which had been thought to be extinct and which became resistant to conventional treatment with antibiotics. Just when we thought we saw the last of some infectious diseases — medical research teams: back to the drawing board!

If that is not enough, new viruses pop up all the time: one of them is the "Hanta virus" giving the patient flu-like symptoms at first, and often with fatal results. Blinded religious believers would attribute new viruses to "God's" will or to punishment for sinners. Again, the question may be asked: why are ALL people subject to infection and possible death regardless of faith or character, even the Pope may get an infection if he is not careful?

Again, the facts are overwhelming, how human lives and those of all larger and small living creatures are controlled by micro organisms — such as bacteria, viruses, antibodies, hormones, DNA, living cell tissue and by sources of radiation, (terrestrial, solar or interstellar), or by chemical reactions. Every death other than accidental, can be traced to some sort of lost battle between good and bad microorganisms, or any of the other causes from above. High cholesterol levels, tar from tobacco smoke, asbestos dust, cancer cells, lack of minerals and vitamins, carbon monoxide, all can be included in this category of potential and actual causes of death. I cannot emphasize it enough — it is not religious faith that determines our destiny, the proof is overwhelming.

Although scientists are discovering more vital information about viruses all the time, a lot more is to be learned. Considering a virus's size and simplicity as an infectious agent, (no more than a collection of genes wrapped in a protein case), it can cause tremendous damage to all living host creatures. Unlike other living single or multiple cellular structures, the virus does not eat or propel itself. It can only reproduce with the aid of another living cell, but viruses CAN reprogram the cells of the "host" organism.

Some viruses attack with a viscous speed, such as the Ebola virus which has an incubation period of just five to eight days, and a mortality rate as high as 90%. The Ebola virus currently has scientists puzzled, since only one of two known identical appearing strains is fatal to humans. The "Zaire" strain had been fatal to monkeys and humans, while the "Reston" strain appears deadly for monkeys only.

In contrast, HIV (AIDS virus) can incubate for up to a decade or longer, and other viruses, such as herpes simplex, co-exist very well with the host organism and the victim is often unaware of it's presence. Again, ancient religion has not taught us or warned us about the effects of viruses and of harmful bacteria, unless one considers the religious prohibition of pork and shell fish as a primitive warning. Other than that, humanity found the answers the hard way, through trial and error, or better known through "years of painstaking research."

Unfortunately until now, at the end of the twentieth century, we have not yet discovered ALL degenerative microscopic or invisible (radiation) sources causing the inevitable death of all living organisms. The answer and the cause is all in the laws of nature, we have to keep looking and we will hopefully sooner than later find all the solutions. Nature has all the answers (antidotes) waiting and they are only to be found by modern science. Nature appears to have for every "action" a "reaction." All in all, it is still truly remarkable how modern sciences have been able to extend our life span significantly over the past 100 years only, and completely without help from religion.

The Universe as part of nature is based on facts only from the present and the past, known to mankind now or to be discovered sooner or later. Life really is a "lottery" with so many possible ways of being born and dying through circumstances. Birth itself is by chance, depending where and when a person or a creature is born.

NATURE: SELF-REGULATING, ALWAYS TRYING TO CATCH UP WITH ITSELF, YET FOREVER IMPERFECT

Nature is by no means perfect, quite the opposite. However, nature IS continuously trying to catch up to adverse conditions through evolutionary changes. Although it operates in a predictable manner, life is created and destroyed constantly by either "normal" circumstances or by excessive, often predictable events. Just as our Sun and terrestrial microorganisms are responsible for the existence of all life on Earth, solar power and microbes can damage or destroy life. Fortunately to life on Earth, the positive effects of solar radiation have outweighed the negative. Earth would look like the planet Venus today had the Sun's negative effects taken over here as they had on Venus.

Every child has been taught at one time or another that too much exposure to solar radiation causes skin cancer. If we look into the Sun for just a split second, our eyes may be permanently damaged. We know it, but many still insist on ignoring these facts. We also tend to forget, it is either too much or to little solar radiation that not only causes serious damage to microscopic body cells, it also changes the entire surface of the planet, since the

Sun alone is responsible for nearly all meteorological changes, good or bad.

We have arctic and antarctic regions on Earth because of the lack of solar radiation at those points, and we have deserts due to excessive solar heat. Too much solar radiation to the polar caps would at first cause massive flooding from melting glaciers and ice flows. And later, the excessive heat would dry out the planet and eventually evaporate the oceans. Chances are, this is what happened to Mars and possibly Venus many millions of years ago and during an equally long process. All are natural events, predictable by scientists as part of the evolution of the universe.

Nature's waste and inefficiency is most evident in the creation of life. Billions of sperm cells take a run at it to fertilize ONE egg cell. Plants drop thousands of seeds to have only a few seedlings developing from some of them, and most of those are eaten immediately by animal "predators," either as a seed or as a seedling. Many egg-laying creatures have to lay hundreds or thousands of eggs, only to lose a great number of them to predators or to decay before they hatch. It goes on after the remaining hatchlings push through their eggshells, the newborn animal will be fighting for survival from the moment of birth. Only very few survive to reproduce again, and the whole cycle starts all over. Why would a super intelligent, "loving God" want this cycle of waste and cruelty to continue on forever? As an example to the contrary and fortunate for potential prey, the giant crocodiles of Tanzania are some of the most efficient eaters. They only need half of their body weight in food for an entire year!

As believers or non-believers know, only the fittest will survive in nature, over millions of years and thousands of generations. As pointed out earlier, all life-forms must adapt to the constantly changing planetary environment in addition to trying to escape capture by predators. If the changes happen too fast or too drastic, many life-forms will not make it. Man-made pollution and senseless hunting and poaching, as well as the destruction of rain forests are only some of such drastic changes, too fast for most life-forms to adapt through normal evolution.

Had some humanoid aliens been stranded on Earth 300 million years ago after arriving from a planet with conditions similar the Earth today, they would in all probability NOT have survived. Mammals are too delicate to survive in the hostile environment of Earth, as is was at that time. The Earth's atmosphere and weather pattern, hundreds of millions of years ago, were far too turbulent for most delicate creatures to survive. Today's meteorological disasters, the frequency and intensity of earthquakes and volcanic eruptions, are a mere ripple to the system, compared to the dramatic changes to this planet, 100,000,000 to 200,000,000 years ago.

Really! Why would "God" have created this kind of wasteful, imperfect nature? Why on a planet, which was totally inhospitable to life, other than for bacteria and single cell creatures in the oceans for hundreds of millions of years? Believers would say: "God has mysterious ways", as an easy-way-out answer to a logical, fundamental question. Even now, vast areas of our planet are extremely inhospitable to human life or to other delicate land mammals. It is either too hot, too dry, or too cold in so many places, and about 75% of Earth is covered by water.

Why would a "Creator" want millions of creatures to be born, only to be eaten immediately at birth by predators, who will in all probability also be eaten by another predator? Why do all life forms HAVE TO DIE, sometimes in a matter of minutes after birth, or with a lot of luck after hundreds of years? The known fact is, the universe and nature are predictable, with firm laws which cannot be altered through religious belief or superstition. Humanity has the intelligence today to understand and to foresee most natural evolutionary events and should prepare accordingly.

In a hypothetical near "perfect" nature, the birth rate of any creature would be extremely low, possibly only to replace victims of accidents and natural disasters.

If all animals were plant eaters, the only causes of death left to animal life forms would be strictly accidental through collision, fire, weather etc.. Without death through predators, sickness or old age, life could go on almost forever for most living organisms.

Sharks may have almost reached that point in nature, they not only seem to age so little that we cannot really tell a shark's age. They also have no predators other than man and develop very few diseases, if any. This is a near perfect creature with a very low birth rate, as can be expected. The "great whites" are thought to reach an age of 70 to 80 years, but scientists have great difficulty to verify this theory with any accuracy.

More optimistic researchers see no reason why the "great whites" could not reach an age of hundreds of years. They DO know however, this perfect eating machine's appearance has not changed for possibly as long as 400,000,000 years after the great changes to the Earth's surface had slowed down. Sharks have been around long before dinosaurs and other larger land animals. They display a much greater natural resistance against most forms of diseases, including cancerous tumors. However, our habit of polluting the seas may also harm the hardy sharks as they are absorbing dangerous doses of cancer causing chemicals through their prey.

Ironically, humanity today has a great advantage over all other life on Earth, we KNOW and understand many of nature's laws, and life to us is to a certain extend predictable. We may be the only creature, which actually knows that all life forms sooner or later have to die. We understand our chances of survival, our approximate life span and that of most other animals and plants. We have the knowledge to prolong life and to prevent or cure many diseases. Or maybe the rest of the animal world is better off *not* knowing, if, when and how the end is coming? In the end everything is really left to chance, often in a most inefficient, cruel, but predictable manner. Anyone could die a violent death tomorrow or live to be 100 years and die peacefully in their sleep. Man will probably never be able to predict the future accurately, no more than other mammals can.

The facts are clear, life and death are natural events caused by a combination of circumstances. Only during the past decades have scientists been able to understand and statistically predict, prevent or change some of these circumstances. It was not a scripture which taught humanity how to prolong life and how to avoid the effects of cancer causing excessive solar radiation or man-made pollution. It is all part of the great gamble of nature and the survival of the fittest.

To emphasize it again, in the end, human and animal antibodies and body organs still lose the life long battle between life and approaching death, between "good" and "bad" microorganisms. But the time will come — and as sure as we have found antidotes for various diseases — science will find more cures and preventative measures. The great number of other members of the animal family, with a much longer life span than our own, could possible "teach" us the secret to longer and healthier life. There must be SOMETHING we can learn from them?

Crocodiles for example are sometimes said to live to a ripe old age of up to 100 years and the entire species has survived from the age of dinosaurs 60 to 100 million years ago. One hundred calendar years in a crocodile's life span translates into over 200 (human) years, since they reach sexual maturity at about six to seven years of age.

I better explain, since I tend to follow a theory based on human puberty of about 12 years of age, multiplied by 6.5 equals a life span of 78 years. Since crocodiles reach puberty at six years of age, about half that of man's, 100 calendar years are really 200 crocodile years, or 16 times the age of puberty, converted into calendar years. I believe, the calculation of pet cats or a dog's real age is based on a similar formula, since it is said that a cat of fifteen is really in her eighties.

Getting back to the crocodiles, 100 "calendar" years or

16 times puberty is simply amazing, considering their main diet is a high cholesterol source of raw, mostly red meat ! The giant tortoise has the crocodiles beat by a long shot with a life expectancy of around 300 years, and even the small limbless lizard, the "European blind worm" can reach the age of 50. And all this without a balanced diet, without aerobics and vitamin supplements . . . By the way, 16 times puberty for humans would mean a human life span of over 190 years, lets not even think about the giant tortoise!

It is interesting to note how similar the life span of elephants is compared to modern man. After sexual maturity at 14 - 15 years, elephants usually reach the age of 70, all very much like our own life cycle. Except we require a balanced diet, occasional drugs to cure illnesses, and we may need medical surgery to reach age 70. A few hundred years ago without modern medicine and little scientific knowledge, our life span was considerably shorter than that of elephants.

Another well known mammal reaches a very ripe old age. The Orca (killer whale) is known to live up to 100 years and reaches sexual maturity at between eight and sixteen years. Based on the age of puberty again, the Orca would reach an age of about 120 human years very easily. Again, all on a diet of raw meat (seals), fish and squid. Here is another mammal of physical qualities far superior to those of humans.

Trees in British Columbia and Oregon have been found to be as old as 2000 years. Trees in other parts of the world have survived over 4000 years. Plants have the additional advantage over most animal life, parts of them can break off or be cut off, only to grow anew and sometimes even better. Very few animal have that ability, one of them being the Octopus; it can re-grow a lost arm and lizards shed their tail, when a predator attacks them, only to re-grow it also.

Just as many other warm blooded mammals, bats have the ability to hibernate. Their heart rate drops to one twentieth of the normal rate and they can survive on the energy of one third of a gram of fat all winter. It is another marvellous body function of a fellow mammal, which mankind cannot even dream of duplicating. One of our closest relatives, the chimpanzee has four times the strength of the average human, pound for pound of body weight.

Using logic based on those facts, it seems that in order to develop near perfect health and a greatly expanded life span, an entire species has to evolve for many millions of years first. Perhaps perfection comes with age in nature? Perhaps, when humans have been around for a few hundred million years, they will also live three hundred healthy years on whatever junk food is popular by than? In any case, one cannot find a contributing factor to old age and good health from a superior being, a "God." In determining the life span of any creature, there are obvious scientific reasons involved. Evolution for one seems to have a prime influence over it.

WORLD RELIGIONS, TRADITIONS, FACTS AND LOGIC

One well known REALITY about religious faith are the amazing HIGH NUMBERS of major, contradictory world religions. As pointed out earlier, during the past two to three thousand years every major modern religion, as well as minor religious sects, declared themselves to be the only CORRECT faith. Also, they usually consider followers of other believes to be wrong, evil and misinformed. Yet, the Judea-Christian-Islamic religions originate from the same ancient scrolls, only various interpretations caused the separation over a period of many centuries.

Let us now look at the origin of various major religions, and one important point comes to light immediately: all of them were founded between 1500 to 3500 years ago. Buddhism was founded about 525 B.C. Islam is "young since it dates back to 622 A.D. And Hinduism was formed circa 1500 B.C. All the teachings, right or wrong, come from a time without medical research, electricity, modern science, and little knowledge of the universe." Our ancestors had no knowledge about the existence of other continents and their inhabitants. They were unable to understand each others languages at first, when they were met by other invaders or visitors. The necessity of time zones was unknown, and not mentioned in scriptures.

BUDDHISM with over 300,000,000 followers is one of several major world religions. This faith is predominantly practiced in Asian countries in a similar manner as most other major religions — it is based in principle on ancient prophecies and teachings. As all other major religious groups, the Buddhist faith was quite acceptable and even logical 2500 years ago. This religion follows strict adherence to the four "Noble Truths:" Earthly existence is sorrowful and impermanent, (quite correct even now to a point, except life is NOT usually sorrowful), and the "roots of existence are desire and ignorance, leading to rebirth."

Of course, we still have no evidence of rebirth and science cannot prove it one way or the other. "Believing" it is the only alternative and may give millions happiness and spiritual fulfillment. Followers of *other* major religions will not accept this view, since THEY inherited and believe in a DIFFERENT faith. Buddhism teaches that the effects of a person's thoughts and actions in previous lives determines the events of future lives. Again, it may be acceptable to the believer who is satisfied which ancient teachings — in preference over scientific reality and logical conclusions in the absence of facts.

CHRISTIANITY and JUDAISM: reference to both religions has been made throughout the book and I will not go into the history of both religious faiths at this point.

ISLAM may not be quite as familiar to the population of the "western world," but with over 800,000,000 followers it is the second largest world religion next to Christianity. Strictly believing in one (also Judea-Christian) God, "Allah," Muslims submit to his will as the "all-powerful, just and merciful creator." One of the five essential duties of a Muslim is to give a share of one's wealth to the poor. A very noble and logical duty indeed.

However, and similar to Christian belief, according to the Koran (the Islamic Holy Book), there will be a final judgement when the faithful will be rewarded in "heaven" and the evil infidels will "rot" in "hell." When Mohammed "received his revelations" nearly 1400 years ago, this prophecy may have been quite effective with the masses of that time as a kind of "scare tactic" preventing many from continuing to commit more evil "sins."

Thomas Paine had the right ideas 200 years ago: "When Moses told the Children of Israel that he received the two tables of the commandments from the hands of God, they were not obliged to believe him, because they had no other authority for it than his telling them so; and I have no other authority for it than some historian telling me so." He continues: "The commandments carry no internal evidence of divinity with them; they contain some good moral precepts, such as any man qualified to be a lawgiver, or a legislator, could produce himself, without having recourse to supernatural intervention."

He did not limit himself to the Judea-Christian faiths only: "When I am told that the Koran was written in heaven and brought to Mahomet by an angel, the account comes too near the same kind of hearsay evidence and second-hand authority as the former. I did not see the angel myself and, therefore, I have a right not to believe it."

As mentioned earlier, to this day science has not been able to prove a "final judgement" or for that matter, the existence of a "heaven" or a "hell" and what about "a just and merciful creator?"

HINDUISM started to develop mainly in India as early as four thousand years ago and the teachings cannot be traced to one individual founder or prophet. Hinduism has a relatively great tolerance towards other religions, with exceptions of course. Based on the "Vedas," a collection of poetic utterances with reflections upon sacrificial ritual, the Hindu traditional faith promotes the idea of birth, death and rebirth as a complete cycle. As in Buddhism, one's actions today will shape one's destiny in "future lives." It is a noble idea but unfortunately, we have no scientific proof of future "life after death."

Hinduism may have come closest to reality since this religion actually accepts some modern scientific concepts. The universe for example was always thought to be much older than what other religions taught, the cosmos is infinite and birth and death are closely related, according to Hinduism. For centuries, Hindus correctly estimated mankind's existence to be 4.32 *million* years and the approximate age of Earth to be 4.32 *billion* years. In Europe in comparison, during the year 1600 Giordano Bruno was burned alive when he claimed that the universe is infinite and that many other worlds similar to Earth exist. His worst "Offense" against CATHOLIC teachings was his assumption that the inhabitants of these worlds live in peace without an inquisition. Unfortunately since he was not living in a Hindu society, it cost him his life.

A CRITICAL LOOK AT MAJOR SCRIPTURES AND AT RELIGIOUS PROPHECIES OF DOOM AND GLOOM

Considering the circumstances at the time when most major religions were introduced, it is understandable that Islam and Judaism for example taught their followers not to eat pork or shellfish. Without refrigerators or freezers it could have been deadly, particularly in the hot climate of the Middle East. The Islamic practice of giving a regular portion of one's possessions to charity was applicable then and really should be practiced today by followers of ALL religions AND by non-believers. I referred to it earlier, but I want to emphasize it again: the commandment, "Thou shalt not kill" was as important a guideline then, as it should be today. So is the commandment "thou shalt not steal," as theft was and still is a very serious crime in the Islamic faith. A good idea, since it appears to be based on common sense, rather than on religious mythology.

The "Introduction" pages of a modern translation of the Koran, the Holy Book of Islam, states: "Mohammed . . . firmly BELIEVED that he was a messenger of God [Allah], sent forth to confirm previous scriptures. God has revealed HIS will to the Jews and the Christians through chosen apostles." The important point is: Mohammed's "firm belief" and the lack of evidence to turn this "belief" into a proven "fact." Only one man about 1380 years ago, had a strong belief and probably with good intentions. As a result of this ancient belief hundreds of millions of followers of the Islamic faith accept this one man's belief to this day as "truth" and undisputable "fact." Wars are still being fought over it against members of other, sometimes "related" faiths, i.e. Christians and Jews.

Much of what Mohammed was to have said according to the Koran, may have actually been for the better of mankind then, nearly 1400 years ago. If it can be combined with logic, those who accept it may still be helped today. Those of his preachings, which do not apply to the twentieth century and which could harm other human beings, should be treated with the greatest scepticism and care before any decisions are made in compliance with the strict "laws" of the Koran.

Two hundred years ago, when Thomas Paine wrote *The Age of Reason*, he already seriously questioned the validity and common sense of the Bible. Following are more excerpts from *The Age of Reason*, once again emphasizing that logic had been applied by a few selected individuals even then and when the church was much more powerful than during the past decades. The question is, WHEN will the majority of the Judea-Christian-Islamic believers start to separate fact from fiction and live according to common sense?

Here it is in Thomas Paine's words: "It has often been said that anything may be proved from the Bible, but before anything can be admitted as proved by the Bible, the Bible itself must be proved to be true; for if the Bible is not true, or the truth of it be doubtful, it ceases to have authority, and cannot be admitted as proof of anything."

"It has the practice of all . . . Christian priests and preachers, to impose the Bible on the world as a mass of truth and as the word of God; they have wrangled and disputed and anathematized [cursed] each other about supposable meaning of particular parts and passages therein."

"In that [of the chapter] of Exodus the reason given for keeping the seventh day is, because (says the commandment) God made the heavens and the earth in six days and rested on the seventh; but in that of Deuteronomy, the reason given is that it was the day on which the children of Israel came out of Egypt and therefore . . . the Lord thy God commanded thee to keep the sabbath day. This makes no mention of Creation."

"The laws of Moses . . . among which inhuman and brutal law, chapter XXI, verses 18 through 21, which authorizes parents . . . to bring their own children to have them stoned to death for what is pleased to call stubbornness." In the next sentence Paine refers to chapter XXV, verse 4, "thou shalt not muzzle the ox when he treadeth out of the corn." Yet parents were allowed to have their children stoned to death!

Thomas Paine continues: "When we read in the books ascribed to Moses, Joshua, etc. that they (the Israelites) came by stealth upon whole nations of people, who as history itself shows, had given them no offense; that they put all those nations to the sword; that they spared neither age nor infancy, that they utterly destroyed men, women and children; that they left not a soul to breathe — expressions that are repeated over and over again in those books . . . The Bible tells us that those assassinations were done by the express command of God."

Questioning the authenticity of the holy books he says: "Neither can there be anonymous testimony, more especially as to the things naturally incredible, such as that of talking to God face to face, or that of the sun and the moon standing still at the command of a man." In that respect Thomas Paine writes in another chapter of *The Age of Reason*: "The tale of the sun standing still upon Mt. Gibeon and the moon in the valley of Ajalon, is one of the fables that detects itself. Such a circumstance could not have happened without being known all over the world. One half [of the population] would have wondered why the sun did not rise, and the other why it did not set; and the tradition of it would be universal, whereas there is not a nation in the world that knows anything about it. But why must the moon stand still?"

The authors comment to it is that even if the "sun had stood still" upon one man's command, since the moon orbits around the earth independently of the sun, it would have continued on its orbit around the earth.

Finally from *The Age of Reason*: "Each of those churches show certain books, which they call revelation, or the word of God. The Jews say that their word of God was given by God to Moses, face to face; the Christians say that their word of God came by divine inspiration, and the Turks say that their word of God (Koran) was brought by an angel from heaven. Each of those churches accuses the other of unbelief; and for my own part, I disbelieve them all."

"I here close the subject of the Old Testament and the New. The evidence I have produced to prove them forgeries is extracted form the books themselves and acts like a two-edged sword either way. If the evidence be denied, the authenticity of the scriptures is denied with it, for it is scripture evidence . . . I have done no more than extracted the evidence from the confused mass of matter with which it is mixed, and arranged that evidence in a point of light to be clearly seen and easily comprehended; and having done this, I leave the reader to judge for himself, as I have judged for myself." All I can say, there must have been really good salesmen amongst churchmen over the past centuries to keep all the biblical mythical misinformation alive amongst intelligent and civilized nations.

When the Old Testament was written, the lack of scientific knowledge was obvious. In order to "make" people of the day accept the Bible and the Koran, both scriptures are full of warnings and threats such as: "Fear Him who created you" or: "Call on no other God besides God, lest you incur his punishment," under "The Cow", page one, first line of the Koran: "This book is not to be doubted. It is a guide for the righteous." If an elected politician today would try to scare a large group of people to accept his own "strong belief" he would immediately be labelled a "dictator," even if his intentions were basically honorable. Yet, even on the back cover of the translation of the Koran, it is clearly stated: "Universally accepted by Muslims to be the INFALLIBLE WORD OF GOD as revealed to Mohammed by the Angel Gabriel."

Quoting another threat AND prophecy [98:1 98:8]: "The unbelievers among the People of the Book [Jews and Christians] and the pagans shall burn forever in the fire of hell. They are the vilest of all creatures. But of all creatures those that embrace the Faith and do good works are the noblest. Their reward, in their Lord's presence shall be the gardens of Eden, gardens watered by running streams, where they dwell forever."

To this day, neither one of these prophecies have turned into reality. Good and evil persons, believers and non-believers have been and are suffering in agony through illness and all die sooner or later in accordance with the laws of nature. Innocent children are born into living "hell" and suffer through their short life, in Bosnia, Somalia, Afghanistan, Rwanda and the Sudan, in Islamic AND Christian nations. Some of the "vilest of creatures" are born into healthy and wealthy families, where the daily bread is no problem, and they die a peaceful death after a very long life of self-indulgence.

Let us look at some of the Christian prophecies, taken from the Revised Standard Version of the Bible, of which millions of copies have been printed and are usually given away to poorly educated natives of underdeveloped countries, where little is known about scientific reality. When the prophecies of doom and gloom were passed from generation to generation, it was under the threat "do not question the Holy Scripture, what is written is sacred."

Taking one prophecy at random from "Ezekiel" 38:15-16, 650 B.C.: "and come from your place out of the uttermost parts of the north, you and many peoples with you, all of them riding on horses, a great host, a mighty army; you will come up against my people Israel, like a cloud covering the land. In the latter days I will bring you against my land, that the nations may know me, when through you, O God, I vindicate my holiness before their eyes." The Israeli general Moshe Dayan in 1968 interpreted it simply as "the next war will not be with the Arabs, but with the Russians." Today, we know how wrong he was, fortunately!

Elsewhere, this quote is explained in detail that the "Russians," (the invaders — presumably meaning the Soviets in those days) would come from the "uttermost north" to invade Israel. It was interpreted that a likely ally of the USSR would be Persia (Iran) . . . so far all wrong and by now expired prophecies, which alarmed people at the time. Of course had the Soviet army really invaded Israel, they would not have "all come riding on horses." We know what really happened to the USSR in the past few years, and if anything, the only invasion of Israel from Russia are tens of thousands of immigrants.

Biblical predictions of this type caused or increased some of the friction and hostilities between nations, in the assumption "if the Holy Book teaches it, it must be right." The timing of this inevitable invasion of Israel was given as "the latter days," apparently during the 1970s and 80s. It was climaxed by the second advent of Jesus Christ, who will come this time as the reigning Messiah. Well, the eighties have come and gone and as with so many ancient religious prophecies, twentieths century reality did not even come close to the ancient religious prophecies of doom and gloom.

The biblical prophet Joel was to have said in 400 B.C. that the Israelites could recover from the plague of locusts by praying and fasting. The obvious question is: did it help? The answer is also simple — it probably didn't and it would certainly not help today either. It has been tried by victims of locust invasions in Africa, always without success.

Maybe not a "prophecy," but a very dangerous threat to humanity in the times of the Old Testament was the Hebrew belief that illness was not to be treated, since it was "God's will for punishment" for evil and sin committed by the victims. It was also believed that dead bodies should not be tampered with, resulting in serious lack of medical research on human bodies. In contrast "non-believers," mainly the Egyptians and the Greeks of the day were clearly more advanced and were able to perform sometimes amazing and advanced surgery, based on discoveries made during autopsies.

Jeremiah, another biblical Prophet warned about the impending destruction of Jerusalem. At the time, it was an easy prophecy to make thanks to the conditions of near anarchy in the area, with countless small "warrior groups" fighting each other on a regular basis. Sooner or later Jerusalem HAD to be destroyed, just as sure as Mt. Vesuvius and St. Helen will erupt again sooner or later. Another prophet, Zepaniah was to have prophesied that all nations will be doomed for their sins at the day of judgement.

Many nations have come and gone since the biblical days WITHOUT being doomed, and others have been around until now, also without being doomed. The first example would be the (evil) former Soviet Union which came and went without "Doom" and the second example is Syria, which has been around since biblical days, also without being doomed to this day. Of course I forget, "judgement day" is yet to come? Leaving the biblical lands: China was also populated long before the Bible was ever thought of, and it now has a population of over one billion. The first census ever taken was in 1 A.D. in China. The population count was an astounding 57,671,400 — (how does THAT compare with a few million in the biblical land?). All wars, including religious conflicts were particularly frequent during biblical days. Even today one does not have to glorify an ancient biblical prophet for super simple conclusions, anyone can predict wars.

Thomas Paine's views need no comment: "The idea that God sent Jesus Christ to publish, as they say, the glad tidings to all nations, from one end of the earth unto the other, is consistent only with the ignorance of those who knew nothing of the extent of the world, and who believed, as those world-savers believed, and continued to believe for several centuries (and that in contradiction to the discoveries of philosophers and the experience of navigators), that the earth was flat like a trencher, and that man might walk to the end of it."

He points to the same objection as I did in another chapter: "But how was Jesus Christ to make anything known to all nations? He could speak but one language, which was Hebrew, and there are in the world several hundred languages. Scarcely any two nations speak the same language . . . As to translations . . . it is impossible to translate from one language into another, not only without losing a great part of the original . . . Besides all this, the art of printing was wholly unknown at the time Christ lived."

Not a prophecy, but just a reminder how often scientists were unexpectedly contradicting the Bible: contradicting the biblical story of Jashua, the middle eastern (biblical) area described was carbon-dated to 2100 B.C. by scientists. Little was mentioned about it in today's media.

Let me now quote and examine some of the explanations from reference books to the various versions of the Bible, or "Books of the Bible." In compliance with modern reality it is said, "very few people today believe that Adam and Eve existed as actual people." Only a few hundred years ago, a similar written statement would have been condemned by religious fundamentalist leaders. Also the statement leaves it open to "very few people." In other words, are there really still a "few people" who believe in, or worse yet, accept Adam and Eve as FACT?

The explanations continue: "Their story is meant to point out the truth that man alone as the only animal can reason and co-operate in God's plans; and that he alone has the power to choose what he does." Of course, the evidence is contradictory. Higher forms of animals CAN reason, as it has been observed from dolphins' behavior or from detailed studies of chimpanzees. In other words, higher form of animal life do not act INSTINCTIVELY ONLY, no more than "man" is acting by REASONING alone. Reasoning and instinct is applied by ALL. Reference to "instinctive human behavior" is made elsewhere in this book.

I strongly support a theory that human attitude of ignorance towards other animals originates from biblical times. Today, hunting and fishing as SPORT are only two examples, along with the slaughter of animals for luxury fur coats. As I stated elsewhere, some native tribes still look upon ALL creatures as sacred and they kill for food and protective clothes ONLY. I see a lot more logic and wisdom in their legends. No other higher form of animal kills other creatures as a "sport," even members of the carnivorous cat family ignore prey within their immediate neighborhood, if they are not hungry. The also "not-so-intelligent" sharks usually appear to kill to feed only.

The biblical saying, "a wolf in sheep's clothing" may have quite possibly contributed to the near-extinction of the wolf species. After extensive and eye-opening research, scientists have discovered during the past two decades that humanity had the wrong impression about these intelligent mammals. Just one more example of where blind religious belief was put ahead of common sense.

Further on in a (biblical) encyclopedia it is clearly stated that the "Lands of the Bible" covered an area "about the size of the state of Oregon." No mention is made about the approximately one million or so South

American Indians and the 50 or so million Chinese and East Indians. The Bible always concentrates on the biblical Middle Eastern countries and their traditions ONLY and it is written accordingly. I find it most notable that before the Bible was actually written, starting about 800 B.C., "the stories of the old testament were handed down by word of mouth." It is well known and still happens today, when a story is transmitted over and over again by "word of mouth" and after going through a half dozen or so people, it can, and most likely will be distorted far beyond the original occurrence.

After the first hand-written scrolls appeared, the Bible went through so many translations and interpretations that it cannot possibly be the same as it was 3000 years ago. The results are evident, the number of contradictory translations for a staggering number of religious groups speak for themselves. Reference is made to God's "creation of man in his image," yet the closest thing to an illustration of God's image in another publications is a painting from the 18th century, found in an old Bulgarian church.

Based on the Christian belief in the Trinity, the painting illustrates the usual bearded image of a white man [the father], and a second one of a younger man [the son]. Both are dressed in middle eastern robes and a dove represents the "Holy Ghost." Is this what we teach children and adults alike at the dawn of the twenty-first century, before we teach them the scientific FACTS? Of course, this is only *one* of many mythical versions of the image of God.

It is indeed fortunate for the entire population of this planet that most of the religious prophecies did NOT materialize. By chance, and as a result of man-made or natural events, some of them may have come close to reality. But with so many religious prophecies to choose from and with so many possible translations, SOME OF IT HAS to come true. Just as a few gamblers WILL actually win sometimes when they play the tables, and some race car drivers WILL be winners and others WILL have fatal accidents.

Getting back to various interpretations of biblical prophecies, one can find the following: "The conflict will not be limited to the Middle East. The apostle John warns that when these two great forces meet in battle, the greatest shock waves ever to hit Earth will occur." If so, will it be by natural force, or by super weapons? THAT is not explained.

John goes on: "all cities of the nations will be destroyed Revelation 16:19)." How clever to leave out the possibility of "natural force of an earthquake," just in case, there is no real war. However, an earthquake or war will never destroy ALL cities of the nations anyway, history has taught us — Amazing, how millions of believers read AND accept it AS FACT! But then in 1970 a Great War WAS a distinct possibility, and prophesied by the Apostle John one should not question it?

The greatest misinformation and prophecy of doom and gloom still is the factual acceptance of a final great war — "Armageddon" as it is called and which will be triggered one way or another by an invasion of the new state of Israel. "The invaders will be judged by the personal return of the Jewish Messiah, Jesus Christ." Once again, all prophecies concentrate on the relatively small middle eastern population of the day. It is another example of how most biblical prophecies ignore the great civilizations of China, India and South America. Their existence was obviously unknown to the authors of the Bible.

All prophecies, biblical and those by non-religious psychics, must be carefully examined for a possible scientific explanation. Some of them have actually materialized, many others have not, just as a gambler will sometimes win, but usually he looses. Other major events have not been prophesied at all. For example, I fail to find a biblical prophecy on the recent Iran/Iraq military conflict where over one million died. It was to be another "holy war" and most warriors died in the belief that Allah has "called" them and that they would be rewarded through everlasting peace in their imaginary "after-life." A lot of wishful thinking, based on ONE interpretation of a scripture, but totally without logic or any evidence of the dead "living" in eternal peace.

There is no harm in BELIEVING in certain prophecies, but in no way should they be accepted AS FACT unless there is a definite scientific explanation for them. Even in this century, hundreds have committed suicide before an impending (biblical) prophecy of doom and gloom was to have materialized, but it never did in the end.

One could go on with quotes from spiritual publications, but my goal is to concentrate on the facts about natural evolution and the universe, on logical theories, and how life and death are really only determined by microorganisms and circumstances. I do not want to burden the readers with many more biblical contradictions and so-called "prophecies" (of doom and gloom, based on "biblical truth"). But I believe it is necessary to emphasize the importance of careful examination of ALL prophecies, and to stress the point that an amount of scepticism MUST be applied to the centuries old scriptures.

CRIMINAL LAWS — INFLUENCED TOO OFTEN BY RELIGION

Some religions forbid the use of alcohol in entire nations. One can find logic in this ancient rule 2000 years ago, particularly in the hot, Middle Eastern climate. Excessive alcohol consumption is harmful to one's health also today, as it was then. It would have been even more harmful in a hot climate, failing to keep the wine-making equipment clean and sterile. As any modern do-it-yourself wine-maker knows, this problem can be re-enacted, resulting in an undrinkable, unhealthy liquid. We now also understand the results of alcohol abuse much better than our ancestors of two to three thousand years ago.

The Christian teaching to be "fruitful and multiply" may have made sense in an under-populated world 2000

Illustrated: An old French carving of Moses receiving the "Tables of the Law" from "God" . . . Why have so many died since then as a result of "enforcement of these mythical, often illogical Laws?" Why are so many in the twentieth century still suffering and dying because of outdated ancient "Laws?"

years ago when human life expectancy was much shorter. Of course, this teaching will under no circumstances allow birth control or abortion. Therefore, it is totally obsolete at the dawn of the twenty-first century. Safe, medical abortion procedures were unknown when the Holy books were written. Also, it would have been completely illogical to apply birth control two to three thousand years ago in an under-populated society. Most of the few hundred thousand inhabitants of the biblical Middle East died at a ripe "old age" of forty, if they were lucky, and if they were NOT killed during one of many armed conflicts. To be fruitful and to multiply would be quite logical under identical circumstances, even today.

We have more people alive on Earth now than have ever died during recorded history. It is no secret, we have a seriously over-populated planet. In the western industrialized world alone, millions of young, single mothers are unemployed or on welfare, trying to raise often unwanted children. Without consideration for these facts, the Catholic church still looks upon birth control as a "sin" and abortion is totally out of the question, because "it is written." Most under-developed, often predominantly Catholic and overpopulated societies are struggling with basic survival, birth control and abortion should be a priority in their situation.

Since an EMBRYO is NOT a BABY, we are NOT killing babies through abortion, at least not in the early stages of a pregnancy. On the contrary, we should encourage abortion at an early stage of pregnancy in over-populated and economically underprivileged areas of the world and in Western society's big cities' slum areas. That is where most teenaged single mothers with children from previous relationships already struggle to survive, making it often impossible for them to ever find, or to accept employment.

One additional child would only make life so much more complicated for these mothers and their families and the cost to society through increasing numbers of welfare recipients does by far out-weigh the cost of free abortions. Of course, this suggestion is completely against most ancient religions, but it is the ONLY realistic answer to one of our present day social dilemmas. Just as we cannot FORCE abortion upon a young, underprivileged girl, we should NOT STOP her to have a desired abortion. If anything, we should encourage it. Yet, we forbid abortion by law, criminalizing it in some of our "developed" nations. This law, as so many others, is clearly based on religious belief.

Municipalities or states paying the cost of abortions in these cases would not only save the tax payers money in the long run, a number of potential criminals would not be born in the first place. Unfortunately it is no secret, the crime rate is highest in the slum areas of most North American or European cities, where the highest number of welfare recipients live. Their chances to get involved in a life of crime is simply much greater. Children born into that environment are victims of unfortunate "circumstances," with no influence from "God," "heaven" or "hell."

For example, it is nearly impossible for a teenager to escape the slums and to start from "scratch" by attending university and finding a high-tech job "outside". The potential lack of educational opportunities is one more reason to encourage and to pay for abortions in municipalities with the highest number of welfare recipients, and with subsequent minimal employment opportunities.

Religious fundamentalists and anti-abortionists do not provide food, shelter, and employment (later) for unwanted children. They very seldom provide decent shelter for underprivileged pregnant teenagers in the first place. Free abortion for these teens would give them a better chance to find employment, without the burden of an unwanted pregnancy and with another child to care for. Of course, it would be very difficult to overcome traditional religious teachings and to change the old anti-abortion laws, where they still exist.

As another example of outdated religious teachings, the biblical "laws" against incest and prostitution may also have been beneficial to the society of Palestine at the time. We now understand the scientific facts about incest that it CAN actually cause abnormal children. This applies not only to humans, but also to the rest of the animal world.

Dog breeders can verify this statement, when their frequently inbred purebreds literally turn into "mental cases" and sometimes have to be destroyed. However, good or bad, incest is a natural fact in the rest of the animal world, and only a small percentage of primates' off-spring is actually born with a birth defect. Therefore, it should be DISCOURAGED, but NOT criminalized, when two CONSENTING ADULT blood relatives enter into a sexual relationship and into marriage. Frequently, such cases are reported by the news media, with details about the daily devastating problems, which a blood related married couple has in society.

As a result of this religious "criminal" law, there have been a number of cases when individuals have gone into a severe state of depression or committed suicide, once it was discovered that their birth resulted from an incestuous relationship. Or when two people in love, got married, had children, and found out AFTER THE FACT that they were actually blood-related. Only the scriptures "teach" that incest between consenting adults is a most serious sin, and our criminal laws are unfortunately written accordingly.

Until recently, a law in the state of Alabama criminalized inter-racial marriages, obviously based on biblical teachings. To this date and in most states, the general attitude is against such marriages. To start with, the biblical "instructions" are contradictory, when believers are taught at the same time that all humanity originated from Adam and Eve and that all men are created equal. The other side of the coin is the teaching against inter-racial marriages. As I demonstrated earlier, twentieth century science and archaeology give us enough reasons to accept the evolution of mankind from a common ancestor and that the genes of all living races are 99.9 % identical. There is no reason not to accept inter-racial marriages, other than illogical religious and traditional teachings.

The biblical teaching on marriage, "Until death do you part," may have been quite sensible 2000 to 3000 years ago. First of all, our ancestors in those days had a life expectancy of only about thirty to forty years. That is, if they were not killed earlier during one of many local territorial or religious wars. Second, since there were so many battles to be fought during that time in our history, and to be fair to the (hopefully) returning husband, the marriage bond had to be much stronger. In those days, death really was almost exclusively the cause of the end of a marriage.

In our modern times, the phrase "Until death do you part" can no longer be applied to marriage with any degree of sincerity. The facts are known — 30 to 50% of marriages fail and the legal implications of divorce are often disastrous, but always painful and costly. Just one more case, where our criminal laws are still far too heavily based on ancient religion, without application of common sense and no consideration for the reality of the twentieth century.

Women in most nations are now much more independent and can quite easily be self-supporting, whereas women 50 or 100 years ago were usually fully occupied with housework and the raising of children. A divorce then was much more devastating, particularly for the wife. Reality has changed so much during the past few decades and in contrast, our traditional criminal laws (biased by religion) have changed very little, if at all.

The traditional marriage and divorce laws in Islamic nations also have not been adjusted to twentieth century reality in recent years and are even more biased by religious tradition. There is no LOGIC in the Islamic law demanding women to wear veils over their faces and to cover their bodies completely, just to list one more outdated religious tradition, yet still applied in the twentieth century.

If we were to apply logic to marriage laws and by

eliminating religion, we would first of all make marriage much more difficult and at the same time make divorce much easier. To prevent costly and ugly divorce proceedings, a marriage should be based on a firm contract between the two partners with particular reference to "what happens" in case of separation. Division of assets, child custody, acceptance or non-acceptance of extra-marital relationships could all be major logical points of importance for marriage contracts. It should be decided between the two consenting adults before the contract is signed, what they wish to do and what not to do during their marriage, but without interference from obsolete, religious "civil" marriage laws.

We prepare for all eventualities through legal, written agreements at the time of a purchase or sale of a property, before financing an automobile purchase, why not prepare for all possibilities before a marriage? Maybe the signing of the marriage agreement could be the highlight of a civil wedding ceremony?

As long as nobody outside or within the marriage is physically or mentally abused, and all partners are in agreement, the two marriage partners should have their choice of life-style. Chances are, if they can work out a difficult contract to start with and both accept it, a marriage WILL survive much longer than many of them do today, while they are based on the old traditional ways of religion. That includes "living in sin" before marriage. If two individuals cannot work out a tough marriage agreement, maybe one partner or both should not EVER get married. Many painful and costly divorce proceedings would never have taken place. Let us be realistic, after the excitement of the preparations is over and the hoopla and glitter of the one-day ceremony is gone along with the honeymoon, life can be very dull and boring for some during the years to follow. The results are obvious and found in daily newspapers and in the courts of law.

In reality, potentially compatible partners who are in love, and who could work out an acceptable marriage contract, are often NOT allowed to get married simply because their families practice different religions. It is still happening every day at the end of the twentieth century. Muslims are not allowed to get married to Christians or Jews and worse yet — in many places Catholics (Christians) cannot get married to Protestants, (other Christians). The same people were taught that "God" created all of them equal?

Almost all Judea/Christian religions and criminal laws in the respective countries, permit only ONE spouse for marriage. In contrast, although they all believe in the same "God," Islamic law allows a man to take up to four wives. The only condition is that he must be an equal provider for all his wives. Which religion is right and which is wrong? In any case and regardless of religion, if a man CAN afford to keep four wives and he can find four willing partners, there is no harm done to any of the five willing spouses. Of course, the reverse applies, if a women wishes to get married to more than one husband, and provided that all partners enter into an acceptable marriage contract, there is no LOGICAL reason, why this relationship should be criminalized. As it stands, there are only RELIGIOUS reasons.

At any given time, a number of bigamists are serving time in prisons in our modern society. Understandable in a way, since the offender causes psychological harm to some or all of the unwilling and unsuspecting spouses. However, if marriages between multiple partners would be permitted in the Judea/Christian society and all partners were to enter into it willingly, the number of incarcerated bigamists would in all probability be negligible. Legal multiple-partner marriages as a logical solution would make it a criminal act ONLY, if NOT ALL partners were agreeable on the existence of the other(s) and if the marriage contract was for a monogamous relationship in the first place.

Getting back to incest: When according to the Bible, Adam and Eve's only surviving son "took a wife," (wherever she came from) they obviously had children, and the next generation would have clearly had to be from incestuous relationships, (brothers and sisters, parents and children), since there was nobody else around to mate with. In no way is it made clear to the readers of "Holy Books" where this NEW wife came from in the first place. But with only two couples or even six or eight blood-related people to start with, multiplication is possible ONLY through incest. Of course, I must emphasize that forced incest even today (father — daughter etc.) IS and should be criminalized through logic, since another person IS HARMED AGAINST her will.

Prostitution is another matter. When this "oldest profession" was practiced first about 4000 years ago, safe sex was unknown and impossible. Diseases were transmitted due to lack of protection and through uncleanliness. Today, with proper protection and if practiced between consenting adults, the natural power of sexual drive should no longer be criminalized simply because it is based on ancient religious teachings. In the end, it would allow our police forces and the legal profession more time to fight "real" crime.

As the final illustration of modern (religious) criminal laws, let me look at the death penalty for convicted murderers as it still applies in many countries and US states. Until the discovery of DNA and its application to crime, suspects were convicted by a jury or by judges, based on physical evidence or in many cases, on circumstantial evidence only. Through the introduction of DNA testing, many criminals have lately been proven to be guilty. DNA sampling is accepted as undisputable evidence. However, DNA tests CAN work in favor of a falsely convicted murderer, if physical evidence is still available for testing.

In the past few years, a number of American death-row inmates have been successful in having their case re-opened and to prove their innocence through DNA testing. Most recently, for forty-eight of them it was more luck than anything else that they had not already been executed. Their DNA tests proved their innocence beyond a doubt, and all of them have been set free. Aside from the emotional turmoil, which they must have gone through while waiting for their execution on death-row, at least their lives have been spared.

If only one innocent convict is executed — it is one too many. Here we have a situation, where almost forty-eight innocent people nearly lost their lives in the name of justice. It is obvious that an equal number of innocent convicts were actually executed before DNA testing, and because of criminal laws which are based on religious teachings "an eye for an eye." Let me look at more logical, alternative means of punishment in my last chapters.

Of less importance, but nevertheless based on religion is another law still on the books in selected areas: It is the prohibition of Sunday shopping and the existence of "dry counties." After much resistance from religious groups, the law has been reversed in a number of places in recent years, often with the exception of liquor sales on Sundays. In some provinces and states this law was, or still is called the "Lord's Day Act." How much more religiously influenced can a law be?

The point is, at all taxpayers' expense, business owners had been visited by police officers, served a summons and later fined in court. The entire procedure had to be paid by ALL taxpayers, believers and non-believers and regardless if they agreed with the law or not. Based on 3000 year old religious teachings, the law criminalized Sunday shopping by ANYBODY willing to open their shop and customers willing to buy. As long as nobody is FORCED to open shop on Sunday and forced to work, there is no logic in this law. In my opinion, and in that of millions, religious believers of one faith in a democracy have no right to dictate their belief onto everybody else; they have no right to stop willing customers and willing entrepreneurs from conducting business on Sundays.

The controversial law of criminalizing assisted suicide is equally based on religious belief. The best known examples are the trials against Dr. Jack Kevorkian, who has assisted with the suicide of over twenty terminally ill patients. Again, taxpayers of all denominations, agnostics and atheists alike have to fund costly trials, when a licensed physician assists a dying, terminally ill person to end misery through death with dignity. The presence of a "living will," and of at least one more independent witness should be enough assurance that no murder is committed. After all, thousands of terminally ill patients are able to and DO commit suicide without assistance and it is not a crime. Only those who are physically not able to reach for the sleeping pills (or whatever drug is chosen), are putting their helper into a position to commit a "criminal offense." Where is the logic?

We can find some logical, but mostly illogical teachings in all major religions or traditions. The reader may re-call, Hinduism for example has been practiced by a large Asian majority for the past 4000 years, guided by the gods Vishnu and Shiva and the goddess Shakti. The multiple of gods is combined in but one principle and are only an aspect of unity. Therefore, by not believing in "our" God or Allah, does that make Hindus any worse or are they sinners in a Christian sense? Just because millions of Hindus have a somewhat different belief? Of course not. They are fellow human beings, they are our "equally created" brothers and sisters and often live probably closer to the rules of common sense compared to believers of other religions. OTHER religions may even practice the Christian rule, "do unto others as you want them to do unto you," more consistently than Christians ever will.

Hinduism also may teach ancient "beliefs," which may or may not be applicable now, and are not logical after we compare them with today's realities and knowledge. Every religious teaching applies to the DATE of writing, to the environment of the time and the degree of scientific knowledge, compared to what we know today. And to a degree of "common sense," if any.

Without modern scientific knowledge 3000 years ago, mankind NEEDED SPIRITUAL guidance and enlightenment so much more than today. The laws of the land had to be based on religion in the absence of scientific knowledge. It is understandable therefore when traditional religions are no longer satisfactory TODAY and become less acceptable to a great silent majority, resulting in a noticeable decline in orthodox religious belief. This trend is no secret and even strong believers and spiritual leaders have to admit to it. Let us remember, religious faith is an inherited BELIEF, whereas the facts speak for themselves. Unfortunately, many educated people still seek refuge in what they consider religious "truth." They tend to overlook the facts and some may eventually end up as members of a religious cult searching for spiritual fulfillment without realizing the consequences.

Without knowledge of the facts, what choice did the old civilizations have? Fantasy based on a flat Earth as the centre of a small visible universe was easy and simple to accept as "gospel" truth, together with the biblical explanation of the origin of man. Both stories were quite acceptable thirty centuries ago. Our ancestors did not KNOW better. Just as we inherited our genes from them, tradition and religious belief was carried forward for generations. Now, the time is almost overdue to look at the realities and scientific facts of the twentieth century, near the dawn of a new millennium.

Let us look again at the simple reason, WHY an individual believes in one religion or another. Nine times out of ten, it is because the parents and grandparents were brought up with this particular faith, in the same way as THEIR parents and ancestors. Since it was, and still is forbidden to question ancient religious teachings, this family tradition potentially carries on forever, without change or until a modern "rebel" surfaces.

I write this book partly because I was brought up by religiously indifferent and divorced parents. Church was only attended for christenings, weddings and funerals, going through the motions like millions of others do today, yesterday and tomorrow. Fortunately I was always free to think for myself and apply logic and fact first. Had I been born into an Amish Mennonite family, as an example, I would have been afraid to even think about what I write today.

THE REALITY OF STATISTICS AND WORLD RELIGIONS

Unquestioned acceptance of an inherited religious family tradition leaves us to this day with large areas of the world and entire nations, where one religion rules predominantly. Here are some official, approximate numbers as of 1987, as reported in a daily newspaper:

RELIGION	WORLD POPULATION	AREA OF PREDOMINANCE	NUMBER
Christians	1,619,000,000	Europe, America	1,138,000,000
Muslims	840,000,000	Asia, Africa	796,000,000
Hindus	648,000,000	Asia	644,000,000
Buddhists	307,400,000	Asia	306,135,000
Sikhs	16,161,000	Asia	15,898,000
Jews	17,981,000	N. America, Europe	12,700,000
Non-religious	1,026,000,000	Asia	776,000,000
other religions	448,000,000		
Total 1987	4,922,542,000		

"non-religious" includes "atheists." "Other religions" includes the following: 109.5 million "new religionists," 97.5 million "tribal religionists", 13 million Shamanists and 5.6 million Confucians. Remarkable is the high number of self-confessed "non-religious" people, second only to the leading number of "Christians" and running ahead of Muslims.

This statistical proof of HOW localized most religions are, makes it easier to understand why we have a lack of communication and understanding between nations, races and neighboring people. However, I take this particular statistic as somewhat hypothetical — depending on nationality, all religions are practiced regularly ONLY by ten to fifty percent of the recorded followers of any one religious faith.

Of the many people who claim to belong to one religion or another and whom I have met over the years, by far most "confessed" that their practice of religion in a "house of worship" was usually restricted to weddings, funerals and other once-in-a-lifetime occasions. This observation at least applies to North America, Northern and Central Europe, most of Asia and northern nations of Arabian Middle East. Although many BELIEVED in a divine creator, a "God," no-one was able to describe this hypothetical superior being, except from various ancient religious paintings.

For further reference and as a comparison, following are the 1993 figures of religious distribution, with the changes in percentages, over the past six years:

RELIGION	WORLD TOTAL	% CHANGE from 1987
Christians	1,833,000,000	+ 0.5 %
Muslims	971,328,000	+ 0.6 %
Hindus	732,812,000	+ 0.2 %
Buddhists	314,940,000	- 0.5 %
Sikhs	18,800,000	-----
Jews	17,822,000	- 0.1 %
non-religious	1,116,542,000	- 0.4 %
other religions	474,743,000	- 0.4 %
Total world	5,479,987,000	— Population Increase since 1987 about 11 % or 557,445,000

Although the world population has increased by nearly 11% in religious belief [or non-belief] are rather insignificant. No religious leader seems to have been able to convince a greater number of undecided "believers" to become new "followers", with the exception of some minor, often fanatical religious cults.

While the numbers of religious believers have not fluctuated greatly, neither have tense international relationships changed. They have only shifted, often because of religion, from hostilities between large political blocks of nations to small, newly "created" states (former USSR and Yugoslavia), again dominated by rather fanatical religious leaders.

International harmony in the presence of blind religious belief has not been and will NOT be achieved, only common sense will contribute to improvement of this situation. Religion never has, and will not in the future contribute to improved living conditions, to better health and to greater knowledge about the facts and laws of nature. Religion can really be very much compared with politics. Taking one example; during the thirties and during World War II, German children and young adults were indoctrinated about the superiority of the Nazi system. I know, I was there, and I believed what I was taught. Just as a young child in a democratic system believes in what is taught in Bible school, and what the young high school student in the former Soviet Union was taught and believed about the greatness of communism. Or the Iranian child believes what is taught about the "great Satan," the USA.

To be more specific: children and the youth in the former Soviet Union between 1918 and 1990 were taught about the superiority of the Marxist/Leninist system. All dictatorship regimes use similar methods of propaganda, in the same way as fundamentalist religious systems, to get the young school children brainwashed. Political leaders of dictatorships, as well as religious fundamentalist "democratic" leaders do not stop short of convincing the youth to die for their cause. Good twentieth century examples are the Islamic regimes of Iran and Libya.

One can find no real advantages in the lives of the population of nations ruled by religious fanaticism, compared to citizens of less religiously influenced governments. Illnesses, life expectancy and causes of death are basically the same everywhere. On the other hand, the disadvantages are often quite obvious — over one million Iranians and Iraqis died during, what Iran declared to be a "holy" war.

When a child or young person is continuously subjected to one-sided propaganda and at the same time is denied knowledge about ALL the events, he or she usually follows a political ideology or the religion of the masses without question. The same still applies to the youth of the People's Republic of China, and to the still communist nation of Cuba, where it is not religion, but where communist ideology is still taught in schools.

FACTS ABOUT DEATH AND DESTRUCTION IN THE NAME OF RELIGIOUS FAITH

As a start of this chapter, I will introduce relevant quotes from Thomas Paine's *The Age of Reason*, including a biblical quote: "Besides, the character of Moses, as stated in the Bible, is the most horrid that can be imagined. If those accounts be true, he was the wretch that first began and carried on wars on the score or on the pretence of religion; and under that mask, or that infatuation, committed the most unexampled atrocities that are found in the history of any nation, of which I will only state one instance."

"When the Jewish army returned from one of their plundering and murdering excursions, the account goes as follows: Numbers, chapter XXXI, verse 13: And Moses, and the Eleazar the priest, and all the princes of the congregation, went forth to meet them without the camp; and Moses was wroth with the officers of the host, with the captains over thousands, and captains over hundreds, which came from the battle; and Moses said unto them, *Have ye saved all the women alive*? Behold, these caused the children of Israel, through the council of Balaam, to commit trespass against the Lord in the mater of Peor, and there was a plague among the congregation of the Lord. Now, therefore, kill every male among the little ones, and kill every woman that hath known a man by lying with him; but all the women-children, that have not known a man by lying with him, keep alive for yourselves."

Paine's quote continues: "Among the detestable villains that in any period of the world have disgraced the name of man, it is impossible to find a greater than Moses, if this account is true. Here is an order to butcher the boys, to massacre the mothers and debauch the daughters. Let any mother put herself in the situation of those mothers; one child murdered, another destined to violation, and herself in the hands of an executioner; let any daughter put herself in the situation of those daughters, destined as a prey to the murderers of a mother and a brother, and what will be their feelings? It is in vain that we attempt to impose upon nature, for nature will have her course, and the religion that tortures all her social ties is a false religion."

Was THAT the first recorded "ethnic cleansing" through mass murder on the principle "If you do not believe MY religion you must die?"

Without modern means of travel and communication, even as late as 500 years ago, mankind would not have been readily able to find information about distant religious or territorial conflicts. Even if they had heard rumors about anything of the kind happening 2000 miles distant, the story was probably badly distorted and misinterpreted, due to the lack of proper means of communication. No surprise under those circumstances that many thousands of Muslims were killed and tortured by "Christian" crusaders during the nearly 200 years of crusades between 1096 A.D. and about 1270 A.D. Christians back home, if they knew about it, would not have dared to QUESTION these "expeditions." Most of them were unaware of the events happening thousands of miles away. In the end, temporary territorial gains made through these religious conflicts only resulted in understandable hatred and revenge action by the losers in later years. This applied then and still does today. Of course, this observation applies equally to illogical territorial conflicts.

Why in the name of a religious BELIEF, and at the

dawn of the twenty-first century, would any statesman order thousands of his nationals to their death, only to cause death and destruction to other human beings of a DIFFERENT FAITH and in the name of a "holy" war? Is it really only because of their difference in religion? If so, it is beyond all logic and even against one of their own religious commandments: "Thou shalt not kill."

Another example of cold blooded mass-murder in the name of religion were the so-called "Wars of Religion" in France between 1562 and 1598. The list could go on into our century, as we only have to look again at what happened in Lebanon, Northern Ireland, the Middle East and at other conflicts where religion was, and still is the basic contributing factor for the hostilities. Afghanistan and Iran are two more examples for this imagined "reason" for killing and destruction, through so-called "Holy wars" and by "enforced" religion. Of course, once again one must not forget the shameful religious conflict between Bosnia and Serbia, and the "ethnic cleansing," conducted by our Serbian, "Christian brothers."

Although we hear or read little about it, another religious conflict or "ethnic cleansing" has been going on for several years in southern Iraq, almost under the noses of the United Nations peace-keeping forces. For reasons which are unclear to many, Saddam Hussein's Sun-ni Muslim-lead dictatorship has been trying to eliminate the mainly Shi'ite Muslim Marsh Arabs of southern Iraq. There may be more than a religious sectarian conflict involved, but it seems to be the main motivation of the Baghdad dictatorship.

Estimates are that over 300,000 Marsh Arabs have been killed so far, and the destruction, torture and mass-murder is still going on between two Islamic sects. The area is, according to the Bible the location of the "Garden of Eden" and was, until now very fertile land for agriculture. Saddam's government is systematically destroying the land by draining the marsh, after poisoning it first. Yet, we have all seen Saddam Hussein on television, when he was shown during prayers . . . Hypocrites are everywhere, not just in North America or Europe.

The readers may recall Saddam Hussein's attempt to eliminate as many Kurds as possible in northern Iraq, supposedly for mainly political reasons. However, it is quite conceivable that religious motives were also involved here. Over 100,000 were killed, many by chemical weapons. Many readers would still remember the footage on television, when the bodies of women and children were discovered, in what is now "free" Kurdistan. A small modern version of Hitler's Holocaust. And it is barely fifty years later.

Blind obedience to the Bible and other religious traditions does not stop fundamentalists and firm believers from using, or abusing the ancient written word at random and to their own benefit and interpretation. As one example, if "God created all men equal," why did one so-called "Christian" organization, the Klu Klux Klan hate, harass, torture and murder their Christian "brothers" of African origin? How do we explain two biblical contradictions, the so-called "God's chosen people" and "All men are all created equal?" Both statements can be found in the scriptures. In REALITY, all human races ARE basically equal in medical terms and physical appearance.

When a Jehovah's Witness denies his own child a life-saving blood transfusion, it can only be classified as sacrificing a life in the name of a doubtful interpretation of belief, without application of logic and reality to this decision. Surely, the "consumption" of blood 2000 years ago may have been dangerous to one's health It could have had fatal results, considering non-existent refrigeration and non-availability of proper tests for infectious diseases. The technology of blood transfusion was unknown at the time. We end up with another typical example of death as a result of an out-dated, illogical religious belief which had been chosen over knowledge. Granted, a small unfortunate number of patients actually has died as result of contaminated blood received during a transfusion, but millions of lives have been SAVED at the same time.

To sum it up, blind obedience to religious beliefs resulted in far more deaths and destruction and more setbacks in research, than most other man-made or natural events over the past twenty centuries. Volcanic eruptions, floods and earthquakes cannot possibly equal the number of casualties from religious conflicts during the same period of time.

Besides death and destruction caused by religious leaders and blind followers, billions of dollars, time and labor have been wasted on "houses of worship." Priceless domes, churches and mosques have been built over the past centuries. In contrast, a few other religious groups such as the Salvation Army could be admired for building very modest houses of worship. Their priority is commendable, it appears to be in providing shelter and assistance to the homeless and poor first.

Unfortunately, their decisions are also often based on pure religious priorities. Recently in Western Canada, the Salvation Army refused money collected through a specially organized bingo game. It is against their religion to accept money from gambling. On the other hand, some money is regularly donated to the same organizations, funds which may have actually originated from lottery winnings or possibly even from proceeds of crime (smuggling), but the origin of the money is unknown and therefore accepted by the Salvation Army.

Only few other religious minorities practice a similar modesty in building of houses of worship and may be admired for it. In contrast, billions of dollars are annually donated by brain-washed followers to so-called TV evangelists. As we learned from experience, most of the funds are used for their own lavish life style and for the construction of luxury resorts.

A drive through a suburban neighborhood in any North American city will inevitably take the observer past a number of modern, usually extravagant "houses of worship." I am not referring to century-old churches, they have been around a long time — I am looking at ultra-modern, luxury marvels of architecture. Sometimes one can find buildings for two different denominations side by side. The necessary funds are usually collected from the worshippers and the rest is mortgaged through the congregation's donations. A logic compromise would have been a modest, functional building in a commercial zone, built for the homeless or as low-rental housing for the under-privileged with a large, but simple basement "recreation hall" as meeting place for the congregations?

Let us search again for common sense in ancient religious teachings: it has been an old Islamic and Hebrew tradition to have the "ritual," compulsory circumcision for male babies. Without proper hygienic facilities and poor cleanliness 3000 years ago it may have been a good idea at the time to avoid infection. What logical reason is there for it in today? This obviously painful practice serves absolutely no purpose, and a number of male babies are known to have died through complications from circumcision. There is no common sense to be found here, only more victims of blind religious belief.

Mennonite communities build modest churches without contracting high-priced builders. There IS actually LOGIC to be found in this action, along with the Mennonite tradition to help each other rebuilding after a home or a barn had been destroyed by fire or other causes. The entire community helps in rebuilding. Religious belief CAN at times be logical.

Taking another example: Christian organizations such as World Vision and other groups, DO practice what they preach by helping the starving children of under-developed nations. They deserve my full support, as long as they concentrate their efforts on SAVING LIVES, and not on forcing their religion upon the recipients of their aid. Ironically, many victims of malnutrition had to be helped BECAUSE they were also victims of a religious conflict. The starving refugees of the Sudan are the best example, where a Moslem dominated regime in this instance systematically victimizes the Christian population and those of any other religious faith.

In the "Western" world, many shelters for the homeless could have been built by the Roman Catholic church from the fortunes invested in the Vatican and other costly cathedrals around the world. How many low-rental residences for the elderly and hospitals could have been built from the donations collected for the construction of these "houses of worship?"

How many hospitals and medical research projects could have been financed from the proceeds collected by "TV evangelists," during their religious brainwashing sessions in front of hundreds of thousands of believers? How many hungry people around the world could be fed from the donations made to various churches, synagogues and mosques in the industrialized world, money which at present is used for building maintenance and for the construction of bigger and better buildings?

Meanwhile, the Pope and other religious leaders instruct their followers to stop questioning the holy scripture, regardless if it makes sense or not. Only a few selected minority religious groups exercise restraint and their priorities ARE in helping the hungry and the poor. As one of many thousands of donors, I have donated to such an organization, NOT because they happened to be Christian oriented — only because they help starving children around the world.

It would be much more beneficial if theologians would teach the FACTS OF MODERN LIFE in existing churches, synagogues and mosques. They could be the ideal "community halls" to introduce young people to the importance of the few logical religious commandments, such as "Thou shalt not kill." Houses of worship should under no circumstances be used to preach against other religions, or to bless weapons of destruction.

Religious leaders' place in society at the dawn of the twenty-first century could be in educating the youth about the realities of life, such as the illogical tradition of the "right to bear arms." In this aspect, videos of some of the

15,000 or so annual gun-related fatal American victims could be shown in "houses of worship," along with videos of senseless automobile accidents caused by alcohol or drug abuse and when lives were lost. These are only a few examples of true contributions which could do much more for the future of our youth, than teaching them prayers and illogical ancient religious traditions.

Preventative measures for general health (proper diet), sex education and birth control, CPR and mouth-to-mouth resuscitation could be other priority subjects to be taught, along with some of the basic facts and theories of the universe. How many teens or children know what to do in case of fire in their home? How many adults know what to do in this case, for that matter? Why not take a group of children to public swimming pools for swimming instructions instead of "Bible" classes, good young swimmers could save lives, including their own. Churches would be ideal places to teach how to save lives.

Unfortunately, today's standard North American educational systems teach the younger generation disappointingly little about the important realities and the fact-based theories of the universe. The golden rule of common sense and logic: "Do unto others as you would want them to do unto you," should be stressed not only in separate religious schools, it can be applied at all times and by everybody.

Of course, even this rule has exceptions, as all others do and must be treated with common sense. For example, anyone contemplating suicide or anyone practicing masochism cannot be considered a good example, (do upon others.) Or one could apply this rule of life to alcoholism and drug abuse: when an alcoholic accepts alcoholic beverages from others, it does not imply he should do the same "unto others!"

Getting away from religion for a moment but still looking at "logic": Europe made a giant step towards logic and common sense, when the European union was created, thereby almost completely eliminating the danger of future military conflicts in the area. It was quite an achievement, considering this union developed between formerly traditionally hostile nations of different cultures and languages. At the same time, other parts of Europe made two giant steps into a dark and hostile past, when Yugoslavia and the former Soviet Union started to split up into independent republics. Here we have opposing examples of positive and negative human "logic."

While on the subject — to the outside observer, the former USSR had one positive point going for itself — an apparent "Union" of cultures and languages in fifteen republics, a "United States of Eastern Europe and Central Asia." Lately, what happened to the priorities? Where is logic applied? Of course the apparent harmony between cultures in the former Yugoslavia has equally been destroyed by senseless separation based on tradition and religion.

In both cases it has been overlooked that we are all circumstantial inhabitants of this planet, with enough space for all of us, if only we live in harmony and peace and if we use nature intelligently by not "abusing" it. Or at least we must manage to co-exist. And this time I will not blame religion alone for it — human emotions, tradition and protectionism are equally at fault.

It must be quite obvious by now, nobody gains in the end from territorial or religious conflicts. Nobody ever has gained in the past from wars except for a few politicians when they were on the side of the winners. In modern days, what have the Lebanese people gained from years of religious civil war? Tens of thousands lost their lives and almost everybody's life had been affected in a detrimental way. Beirut is destroyed and it will take decades to rebuild. And now the historical cities of Bosnia go through the same senseless experience.

Another religious conflict still lingers on in Afghanistan, with a power struggle between various Muslim factions. At the same time, the conflict in Northern Ireland was nothing to be proud of for Christians. As for gains from war: the second world war, which was not a religious conflict, ironically resulted in the losers, Germany and

Japan, to emerge as the strongest economies in the world. A good place to teach prevention of armed conflicts would be in houses of worship, if and where it is not already being done.

WHAT IS A BELIEVER'S "TRUE" IMAGE OF "GOD" OR "ALLAH?"

At a presidential inauguration or at the start of this ceremony for a new Prime Minister in the UK or Canada, in courts and at other official ceremonies, the following words are usually spoken: ". . . so help me GOD!" If humanity would take religion for what it REALLY is: "Belief," "Faith," and not "FACT," the first logical question to oneself would be: Who is this "God," whom we are asking to help us? Can He be everywhere and help everybody at all times? As long as we have NO facts, we can "believe" that God will help, but we CANNOT ACCEPT IT AS FACT. Since we don't know who or what "God" is, or if his existence is a reality, why do we ask "God" for help during the very realistic and serious ceremony of an inauguration of a head of state? Or in a court of law, during the swearing of an oath? Has a "God" really ever "helped" those who asked for it at these occasions? What did "He" look like?

I have discussed the subject with many religious individuals and non-believers, but nobody has been able to tell me who or what THEIR image of "GOD" really is. They were unable to tell me WHY they actually "BE-LIEVE" in a superior "Creator," or worse yet, why they accept God the person as "FACT." Quite often their only answer was something like that: "There has to be something or somebody to overlook everything, it cannot all just happen by chance."

However, nobody was able to give me at least a vague description of God, and many of them actually discarded the old idea of a "white" God in human forms. After all, NO large group of eyewitnesses actually has ever seen "God" (or "Al-

Fact or myth? Two examples of many ancient Gods: The great monolith of Tiahuanacu in Bolivia, a tenth century statue of the "Great White God" or "Viracocha." The local Indians mistook the invading "Christian" Spaniards for returning "Gods" and named them also "Viracocha." A fatal mistake for most of them.

Second, an eighteenth century painting in a church: The Judea-Christian-Islamic "God" - the "Father," the "Son" (Jesus Christ) and the "Holy Ghost" (The Dove). Are we looking at examples of ancient men of great wisdom, possibly gentle, highly advanced extraterrestrial visitors? Which one of these or other mythical "Gods" is the "REAL" God, if any? There is no scientific proof of any of them having the power to "create" anything, life, the Earth, the stars and least of all, THE UNIVERSE.

lah"). Nobody resembling one of the believers' images, and with the powers of an "Almighty," has ever landed in a modern city square or in Washington in front of the White House. Or for that matter in the Vatican or any other church, mosque or synagogue.

The closest explanation comes from the New Testament when an average normal human being, Jesus Christ, is referred to as the "Son of God." Religious belief, as previously mentioned, is usually only "carried" forward from previous generations, all the way back to the time when the holy books were written, including hatred for believers of other religions.

As for a "Creator," we KNOW the natural FACTS of the "creation" of life. Until a "God" actually appears to thousands of eyewitnesses and if HE can demonstrate the power of creation, accepting God "the creator of the universe" as fact, is much less realistic than accepting the existence of alien visitors. Without SOLID scientific evidence in EITHER case, but with some definite, creditable sightings in the case of aliens, I would be more receptive to the reality of the latter. At least there are SOME acceptable sightings by creditable witnesses of modern times, not just two thousand year old myth.

Most believers claim that "God is everywhere," within all of us. It may have been acceptable myth ten centuries B.C., but no longer in the twentieth century, since we know exactly how big the planet Earth really is. The fact is, we have over five billion humans living on this planet. We understand a great majority of realities of time and space, how could any super-being be EVERYWHERE, ALL OF THE TIME, on this planet and on countless others in our galaxy alone? The laws of the universe, of time, space and matter work everywhere and at all times, nothing and nobody can overcome them. We will look at this subject in more detail later.

After logical consideration of all the options, "who" or "what" God really is, if one wants to go beyond "Belief" in a God to acceptance of Him as "fact," we can again eliminate the traditional image of an old, bearded man. The mythical vision that "God" is "everywhere" can also be eliminated since NO material matter, NO radiation and no other visible or invisible force (i.e. magnetism), can be EVERYWHERE at the SAME TIME. Nothing in the universe has been observed, measured or calculated to overcome the basics of time and space and the nuclear cycle of conversion of matter. Many unsubstantiated theories about "God's" image have been invented by theologians and religious fundamentalists, but nothing has ever been proven.

However, if one considers alien visitors as not only a distinct possibility, and based on thousands of eyewitness reports accepts them as fact, we are left with one possible and somewhat logical explanation of the possible existence of "our" (Christian, Muslim or Hebrew) God: According to native myth and legend, other highly advanced aliens ("A great white god") visited South America possibly during, or before the great period of prosperity of the several millions of Mayans. Other native legends all over the world talk about a visit of some sort of God, or Gods at one time or another to THEIR region. The same may have taken place if and when a technologically highly advanced "Alien" has visited the biblical regions of the Middle East — He (or They) may have had the greatest of intentions for mankind in the region, giving them basic well-meaning "commandments" to live by. With incomprehensible powers and wisdom at THAT time, this "alien" (or aliens — angels?) may have communicated with the locals in a manner and language understandable to the inhabitants of the biblical land about 3,500 years ago.

Believable — yes, but it is not a proven fact. In any event, if this particular hypothetical (alien) "God" communicated with the locals 3,500 years ago, many of his possibly well-meaning warnings and instructions are obsolete today. Without application of logic, they can be harmful to other fellow human beings if misinterpreted. No matter how technologically advanced any alien visitors may appear to be, based on eyewitness reports, they cannot be "everywhere", within "all of us", and they themselves are subject to accidents as the reader can see from another chapter.

Any technologically advanced visitors who set foot on this planet 2000 or 3000 years ago, would have been without doubt immediately worshipped and honored as a "God." Just as many African and Indian tribes greeted any much more recent visitor of superior knowledge and wisdom, so did the inhabitants of the Middle Eastern region at that time. Over time, the wisdom of this visitor or visitors (Angels?) may have been transmitted verbally first and was later recorded on scrolls, etc.

Considering the mentality and almost non-existent scientific knowledge of the locals at the time of this hypothetical visit, much of the information had to be transmitted accordingly and in a manner understandable to the locals — in particular to those who were fortunate enough to meet this visiting alien, (Moses?). In order to introduce the technologically completely uneducated locals to common sense and to the facts of the universe, simple methods and some "parental" and dictatorial methods would have had to be used, along with threats for those who fail to accept the visiting alien's (God's) philosophies.

No doubt, had this visiting alien "God" really existed, he probably really had the best of intentions teaching the population of this small area some common sense and wisdom. Other visiting "Gods" may also have given the population of other lands more or less valuable information for their own good. This is obvious through the absence of signs of violence from any area, which was according to legend visited by advanced aliens in the distant past. On the other hand, evidence of forceful introduction of religion is plentiful, originating from our "Christian" ancestors, the conquistadors and the Crusaders.

As many others before me, I have written about what appears to be evidence of alien "landing sites" in other chapters, but nowhere are signs of violence to humans to be found, originating from extraterrestrial visitors. Interesting enough, we have at least what may be to some questionable evidence of alien landing sites, etc., — Whatever it is, it is a lot more than we have from a Judea-Christian-Islamic God's visit to Earth three thousand years ago.

Although a few so-called encounters of the "fourth kind" were reported to be of an unfriendly nature during painful testing and probing of the abductees, such occurrences seem to be the exception, rather than the rule. In any event, if alien visitors would have had serious hostile intentions, their high-tech weapons, (whatever they are) could have wiped out humanity a long time ago. Yes, "our" mythical God may just have been another far advanced humanoid creature and His "Son" of Christian belief, Jesus Christ may have even been the part-human off-spring of this alien visitor, giving him extraordinary powers. The fact still remains, many of the teachings of the days of "our" God's visit to Earth can no longer apply if they cannot be combined with common sense and scientific knowledge.

Some of the teachings of other visiting "Gods" also do not apply any longer, from the Egyptian "Gods" to those of the ancient Greeks and the Mayans. Today, we do not have to accept dictatorship from any EARTHLY leader, and we should carefully examine 3500 year-old "orders" from a visiting alien(?) super- "human" God. And we should most certainly examine orders and laws from self-appointed religious leaders who order their followers to death and to commit senseless killings, all in the name of one of many "beliefs." That includes following all senseless ancient laws, supposedly in the name of a "God." Let logic and common sense prevail, if no harm is done to any other person or living creature, it cannot be a so-called "Sin."

Most schools, primary and secondary, STILL teach children about "God" the creator, based on the old, familiar picture images of a white-haired, bearded Caucasian male. Our children are taught how God created the Earth and how we all come from Adam and Eve, etc., etc. A lot of precious time out of a young life is wasted on 3500 year old religious myth. Far too much emphasis is given in teaching young children the various religions of their parents, while respect for all living creatures and a certain amount of discipline is ignored. All the while, the teaching of important facts of life is ignored.

Later, the same children are educated in high schools, colleges and universities when they finally learn about the scientific facts of the creation of Earth, about evolution and about the human body, plus the contributing factors that really "make" life and death. What are they to believe? Why not teach children the facts FIRST and leave religion for them to choose as a far-distant SECONDARY option only, and when they are old enough to make a sensible choice? Let's be realistic, "believing" in God and in a religion is one thing, but I repeat it again: To blindly "accept as fact" everything, without sceptical questioning on what is written in various old scriptures about God, Allah or Jehovah, can be, and often is a fatal mistake. It compares to believing blindly in a political movement (Communism), without questioning the realities.

By always using reality and logic FIRST and religious teachings only as the LAST resort, many armed conflicts and much suffering could have been avoided — and still can be today. Just to name one small random example of absolute blind faith in religious teachings, as it is still practiced today: the stoning and lashing of unfaithful wives as punishment for infidelity, ordered by Islamic fundamentalist governments. This little known, but still practiced torturous death sentence, comes out in the open more often today, thanks to modern communications.

Case in point: One very recent newspaper report referred to the public stoning of five women as death penalty, in plain view of United Nations personnel in northern Somalia. At the same time, one unfortunate women was lashed savagely 100 times. She was really "fortunate," at least according to overseeing religious leaders her crime was illicit sex, but she was NOT married while all the others had been married women. UN personnel tried to intervene, but were also threatened with stoning and retaliation if they as much as stepped out of their vehicles. One would think Somalians have more important things to be concerned about!

There is absolutely no common sense or logic in this type of brutal action, and it is one more example how religious traditional dictatorship only knows and applies one law and one law only, in this case, the Koran. Granted, the act of unfaithfulness may by unfair to the spouses and may or may not deserve punishment. But torturing young women to death for it can by no means be considered fair "punishment." Where is the spirit of "Thou shalt not kill" or "do unto others, as you want them to do unto you?" But then, I am applying two of few logical commandments and rules from the BIBLE and not from the Koran. I almost forgot, Christian and Muslim people are not the same, at least not according to individual religious teachings.

By the way, what happens to unfaithful husbands in the same country? Nothing. Why is life not "sacred" for women who happen to be born into the Islamic faith and who are victims of circumstances as a result of it?

Recently on television, we have been shown Muslim refugees fleeing their homes in Azerbaijan from the so-called "good" Christians, the eastern orthodox Armenians. Tens of thousands of non-Christians fled for the Iranian border to find refuge in massive tent camps. The Christian intruders looted and burned the victims' homes and the United Nations ended up with another refugee problem, with thousands of more starving men, women and children. Their only "crime" was to be believers of another religion.

In contrast but on the same principle, many readers may be familiar with the systematic slaughter of Armenians by Muslim Turkish conquerors between 1895 and 1896 and again during 1914 and 1915. Well over one million Armenians lost their lives during this senseless holocaust, about 500,000 ended up as homeless refugees. Again, the religious background is the only separation between Turkish and Armenian people, and hatred persists to this day because of the events of nearly 100 years ago.

Followers of various religions still choose to believe (or accept as fact) that "God" gave us our body and our brain, others accept as fact that it all developed as a result of evolution over millions of years. Let's look at reality again: microorganisms, our environment, our feelings, our instincts and the inherited genes rule most of our life, exactly as the life of other mammals is ruled. At least nature makes no distinction between believers of one religion or another, we are ALL EQUAL under the laws of nature.

PRAYER: WHAT DOES IT DO FOR ANYBODY? IS IT A FORM OF SELF-HYPNOSIS?

To take a simple example: When religious "believers" pray before a meal to thank God for it, they do so basically because their parents and often their teachers taught them to do so. But at the same time, they become blind to the facts. Someone in the family had the education and training to find reasonably well paying employment and his or her income paid for the food on the table. If this same real provider for the family loses the job and can no longer afford to buy enough food, the family becomes a victim of man-made circumstances. Prayer may or may not "help", and even if it (apparently) "helps," circumstances probably have changed for the better anyway, without prayer.

Let us say, when this same individual is extremely persistent and finds a new job after many interviews and a lot of door-knocking, he or she is now a "beneficiary" of man-made circumstances. If it was because of a recession that the job was lost in the first place, the family became a "victim" of circumstances, not a victims of the "forces of evil." After a lot of initiative and hard work, the head of the family finally found a new job and was able to put the "daily bread" on the table. The point is: why should they thank a mythical figure for the food which they now can afford to buy again? The new job was found with or without prayer and regardless of religious belief.

It was all his or her OWN doing, allowing the family to eat better again. And it is the farmers' labor, growing the potatoes, the vegetables and corn and raising the cows, provided it was assisted by meteorologically favorable conditions. By the way, anyone trying to raise cattle in the Arctic or in a desert zone would be unsuccessful no matter what his religious belief is and how much or how little praying and faith was involved.

Let me carry this example one step further: This individual could have done the same as the neighbor when HE lost HIS job. Let us say, the hypothetical neighbor had a lesser education because he had left high school early and on his own accord. In addition, he simply did not have the persistence to keep on trying for another job interview. In any case, when the less ambitious neighbor later gets assistance from the food-bank and he receives social assistance, why should HE pray and thank the "Lord" for the food ?

Our two North American friends and neighbors could be in an entirely different situation if their forefathers had NOT left the "old country" to start a new life for themselves, for their grandchildren and great-grandchildren. What if one of them had NOT immigrated (say from Bosnia) and the family would still be there? If the family of Bosnian heritage still believe in praying before a meal, it would surely be more appropriate to thank their forefathers, rather than a mythical "God?"

There is nothing WRONG in thanking "God" for the daily bread (or for anything else) and for spiritual satisfaction, but it will NOT change reality and the laws of science. Nothing gained, nothing lost. Some believers may gain spiritual fulfillment from prayer. However, their daily "bread" may be gone the next year because of a severe drought. If so, it is because of a complex chain of meteorological events which in turn may possibly be traced back to changes in solar flares or to the massive eruption of a volcano, or a number of other natural events. Not because "God" wanted to punish some "sinners," but as a result of nature's law of probabilities.

No matter how much the inhabitants of the southern part of the Sudan may have prayed, the drought conditions continue to persist and the region effected increases in size and in an almost predictable direction. The only solution is to move away, far enough away to an area where the drought will not hit until much later, if ever. Prayers most certainly did not and do not help the suffering population.

While on the subject "Sudan," here is another classical example of a religious conflict with daily tragic results. As mentioned, the Moslem government-controlled army is involved in a civil war with mostly black Christians and tribal rebels. Foreign food aid can not get to the starving southern tribes and to top it off, a long drought in the area makes it impossible for the inhabitants to remain there. Thousands of men, women and children die daily of starvation as a result of religious conflict AND drought, but not much about it can be found in the western media. Several Christian Aid organizations try to help and should be commended for it, but the odds are against success under the circumstances. Praying certainly does not help and it does not prevent Christian and Muslim brothers from killing each other.

Praying may or may not help anybody in distress, pain, or poverty. Changes to the situation may come about sooner or later anyway, with or without prayers. Even a strong believer must admit, a considerable number of prayers are clearly NOT answered and the results are written off as "God's" will. Our will to acquire a very strong belief is extremely powerful, although this power of the mind can become self-destructing if we believe in the wrong theory or myth.

A strong will or positive thinking CAN result in favorable changes, it is known to even bring cancer into remission. Although mankind does not fully understand the process of mind power as yet, the power of one's thoughts (also known as "visualization") has been quite evident to everybody at one time or another. But one does not have to be religious, the mind alone is doing it. Just as we can learn how to read and write when we are young children, and ultimately graduate from college or University, if we PUT OUR MIND TO IT we can do amazing things!

DID PRAYER REALLY HELP?

When the Tamboro volcano on Java erupted between April 7 and 12, 1815, the entire northern hemisphere experienced a massive deterioration in weather pattern, right through the entire following year. About 12,000 died and millions were more or less adversely effected by it. Buddhists, Moslems, Christians, Jews, non-believers, poor people, rich people, "good" and "bad" guys, all were victims in one way or another. Many believers prayed for relief, but nature took its course, prayers or not.

Just as a result of any other volcanic eruption, the residents of the immediate area either died or were spared, depending on distance from the eruption and wind direction. Some of the land was covered with thick poisonous layers of volcanic ashes and other parts were spared. Not because the victims and survivors near the eruption believed in one religion or another and some prayed more than others.

The facts of the year 1816 are a good example of how natural events effect all life and how little religion and praying can do against it. That year was legendary in weather history. It has been called the poverty year, the year without a summer (Sounds familiar for 1992 with an almost identical reason!). In 1816, between May and September, half-way around the world a series of cold waves chilled New England and southern Quebec causing near-famine in many areas. Between June 6th and 10th, Quebec City was covered by nearly a foot of snow with much higher drifts. Freezing temperatures blackened crops and produced thick ice on lakes and ponds with devastating results.

When Tamboro erupted a year earlier, about 150 million tons of dust, ash and cinders were ejected into the upper atmosphere. It resulted in a massive, temporary climatic change to many parts of the world which are still influenced by the (then unknown) jet streams. It is one more example how natural habitat is subject to cause and effect, how today we can predict the results of volcanic eruptions and of extraterrestrial influences. Thanks to modern science and without any help from religious scripts, prayer did not and will not change nature.

The poor souls in 1816 had little science to rely upon and to prepare for the climatic changes caused a year earlier by this massive volcanic eruption halfway around the world. When their prayers were seemingly "answered" the following year, they most likely thanked God for it. Little did they know that a return to normal weather conditions HAD to be expected, after the dust had settled, literally.

A dormant volcano of modern days, Mt. Pinatubo, 90 km north of Manila, erupted after 600 years of silence on June 9th, 1991. Once again, a perfectly natural event and somewhat predictable — no "angry God" caused 350 "sinners" to die and 100,000 to 200,000 to become homeless. No amount of praying and no matter what their religious belief was, it made no difference to the victims or to the survivors. They all either died or survived as "victims of circumstances." To begin with, they lived in the vicinity of the volcano. Secondly, they probably happened to be inside one of many fragile buildings which HAD to collapse under the weight of the falling debris. Had they been maybe 500 feet away and OUTSIDE the buildings, they may have survived. In a true sense of the word "Victims of gravity."

Residents of California are by now aware of recent scientific findings that a major earthquake will hit the area between now and approximately thirty years from now. This information comes from scientists as a result of years of painstaking research, not from biblical prophecies despite that knowledge, only a relatively small number of Californians have moved to other, less earthquake-prone locations.

When the "big One" hits in a few days or years, the remaining residents are much more likely to be killed or injured as a result of their own desire to remain in the area right next to the St. Andreas fault. When the time comes, prayer will not help before or during the quake. A number of people will again become "victims of circumstances (gravity)", and will have lost the high risk gamble of living in that part of California. Most will probably survive again, but not thanks to religious belief or non-belief, with or without prayers.

The anticipated major Californian earthquake may or may not hit a populated area. When the last most powerful quake hit the West coast, it was in and around Anchorage, Alaska on March 27th, 1964. Since it occurred in a less densely populated area, the death toll was relatively low at only 140. At a magnitude of 9.2 on the Richter scale, it was the most powerful quake in recorded history and it would have killed millions in the highly populated part of southern California.

In contrast, the victims of the great Lisbon earthquake of November 1st, 1755 were not so fortunate when 70,000 lost their lives at that time in a densely populated area. Ironically, it happened in a very religious (Catholic) part of the world.

The unfortunate victims of religious conflicts in Bosnia, Afghanistan and southern Iraq have been praying for peace for decades. Although the hostilities were based on religious doctrine, prayers did absolutely nothing. Victims died and still do, others are injured to a point where they would be better off dead, regardless of their faith. Faithful Muslims die for the same irrational cause, as Christians in the former Yugoslavia. Prayers for peace by Catholics and by Protestants had no more effect on the equally irresponsible conflict in Northern Ireland, one way or the other.

THE GREATEST CLASH BETWEEN REALITY AND MYTHICAL RELIGIOUS BELIEF: SIMPLE NUMBERS

Let us look again at the religious myth of human life being created by an "Almighty God" and that ALL life on Earth is created by God, and each life is ENDED through God's will. Let us not even think about the animals from the smallest insect to highest form of mammals, whose lives would in "reality" also have to be created and terminated by the "Lord." Let us only think about all HUMAN lives for the moment.

Planet Earth currently has a human population of about 5.5 billion (plus). Every SECOND, approximately three human babies are born somewhere on Earth. This instant, while reading these lines, one baby may have been born in Jakarta, Indonesia. Almost at the same moment, another baby may have been born 5000 miles away in Nairobi, Kenya and again at the same moment, a baby may have been born in Mexico City, almost 10,000 miles from Nairobi. Of course these locations are hypothetical, and the numbers may vary somewhere between one and five babies per second.

The fact is, three new human lives appear on an average every second on this planet. Of course, the same numbers apply for conception if that is when life really is "created." In reality, at any one given second, it is quite possible that twenty babies are born somewhere on Earth, while in the next second none are born. Comparatively, some seconds may tick by without a death occurring, while at other times during natural disasters or war, thousands may be killed at the same instant. Eighty Thousand died in an instant in Hiroshima in 1945.

Since there is a surplus of overall birth rates over death rates, one can safely assume that at the same time when three babies are born, rounding up numbers, somewhere on Earth about two human lives have ENDED. Again, the locations may be thousands of miles apart, or the two deaths during the last second may have been two victims of a car crash side by side on a North American or European highway. The point is on average, a total of five lives are effected by birth (conception) or death every single second on this planet. During a plane-crash, maybe 300 people die in the same instant.

Once again, we must come back to the teachings of the scriptures where it is taught, that God has created "man in his image." Therefore, God the "creator", must be at least sometimes human in appearance. His mythical son of Christian teachings, Jesus Christ, was also known to everybody at the time as a young, middle eastern white male of average human appearance.

The logical, scientific question is: No matter how much of a super-human figure the mythical God is believed to be, how could He be in five different places at the SAME TIME, EVERY SECOND for a total of 86,400 seconds every day? Or to make it more difficult yet for those religious believers, who agree with the scientific theory of a minimum of 50,000 probable earth-like inhabited planets in our galaxy: How could Allah possibly "giveth and taketh" life five times every second on 50,000 planets at the same time? That translates into "making and taking" of 250,000 (humanoid) lives per second in our galaxy alone and separated at vast distances of up to 100,000 light-years! Or how could God possibly be at the scene of every accident, sometimes at the same instant "miraculously" saving lives, hundreds and thousands of times every minute; thousands of miles apart and while He is busy "giving" and "taking" lives through natural birth and death?

Let's examine numbers again: Our mythical God would have exactly one fifth of a second to decide the fate of a life or one second to decide on three births and two deaths in totally different locations on Earth. In one day on this planet alone, about 432,000 human lives are either "created" or ended, or 21.6 million lives are created or taken between the hypothetical 50,000 earth-like planets within our galaxy every 24 hours, on planets which are separated by hundreds, thousands and ten thousands of light years.

Up to now, I have primarily concentrated on "giving" and "taking" of HUMAN lives. However, the Bible teaches that animal life is created and terminated by God also. The hypothetical numbers get even more astronomical if we take the totals of ALL "higher" animal life forms in consideration, including our pets alone. To this point I have referred almost exclusively to birth and death only — what about the many instances every minute when believers pray to God to be heard, to thank or to be helped? How could "He" possibly "hear" and respond to them all?

One question comes back again — with "His" hypothetical powers, why has God not appeared to billions of believers on Earth in recent years, not only to teach us the latest facts but also to demonstrate his most improbable powers? No known force in the universe can perform these kinds of miracles, nothing can overcome astronomical distances in an instant. Statistical and scientific numbers are facts which cannot be altered by any known force. By the way, was it not just Moses ALONE who actually saw "God," according to the Bible?

Reported sightings of apparently most technologically advanced alien spacecrafts all seem to agree on the awesome maneuverability and speed of the UFOs. But all of them are also subject to space, time and gravity. They are seen at any given time in ONE particular portion of the sky or on Earth, and seem to move at a relative speed, i.e. they do not appear in an instant from invisibility, or visa versa. They may be travelling at speeds of close to 180,000 miles per second, (near the speed of light and radiation), once they leave the gravitational forces of Earth and the solar system.

With an estimated 40 billion known galaxies in the universe, each one of them with maybe not 50,000 but conservatively with no more than 20,000 inhabited planets, it must be difficult for even the most faithful believer to imagine God to be IN BILLIONS OF PLACES AT THE SAME

TIME. Every second of the day for billions of years and separated by distances of up to billions of light years? Even the theologians' theory that God can do "anything" should get somewhat shaky under the pressure of those numbers!

However, when the commandments and biblical prophecies were written, the authors only knew about their small world the size of Oregon around them with maybe as few as ten thousand or possibly as many as 100,000 inhabitants (there was no census taken in 1000 B.C.). The combined population of Palestine, Egypt, Syria, the Assyrian Empire, Babylonia and Phoenicia as late as the seventh century B.C., may not have reached one million. God's domain was to their knowledge restricted to that area and the theory of creation and termination of life through God's will could have been quite acceptable to those who did not dare to ask questions. But since God "created the Earth, the stars and the universe," all life must be his responsibility. In any case, should it not be the other way around, the universe would have had to be created FIRST?

As noted elsewhere, an almost exclusive reference to the Middle East as the "world" in the Bible completely ignores the fact that other advanced civilizations already existed at a time when God was to have created the Earth. One good example are the Mayans of South-Central America, who by about 300 B.C. already had a population of several million, far more than the reported population of the "Middle East" at about the same time. Elsewhere in this book, I have mentioned other civilizations with an even much older history (China).

HIGHER LIFE FORMS AND THREE BASIC VARIETIES OF "LOVE"

Love for your spouse, loving one's children, love between siblings and love between friends — all are familiar forms of "love." We can add biblical phrases to the "Love Goulash," "Love thy neighbor," "Jesus loves you," etc.

Upon closer examination, one must come to the conclusion that there are really SEVERAL TYPES of "love" relationships. Some are identical to "love" between other animal mates or parental love towards their off-spring. Biblical teachings about love are mostly hypothetical, i.e. would Jesus love a follower of the Hindu religion, Muslim, Atheist? Has everybody experienced His love, regardless of religion? Once again I am examining the impossibility of God or Jesus being everywhere, in contact with everybody at all times. That includes Him "loving" ALL humans.

Sexual intercourse for the purpose of fertilizing female egg cells between OTHER animals of the two opposite sexes is really "MATING." Pure sexual attraction between men and women is based on the same genetically inherited animal instinct. Sometimes it may be called "LUST," but it is often misinterpreted as "LOVE." In reality, it is a natural instinct for the purpose of reproduction and for survival of the (human) species. "Falling in love" is more often than not the result of a sexual, purely physical attraction between opposite sexes, before real "Love" between partners may or may not set in later.

Therefore, sexual relations between consenting adult persons are a natural and instinctive behavior. The Bible and other holy books condemn it as a "sin" to engage in sexual activities, unless a couple has entered into a marriage and intend to "be fruitful and multiply." As a result of biblical inspired laws, we waste millions of dollars in courts and through law-enforcement annually, by arresting and prosecuting consenting adults, (prostitutes and their customers). "Modern" societies actually CREATE crime by turning a natural activity into a criminal offense, and by driving it underground. Elsewhere, women of various faiths are stoned to death over it in the name of the same God.

Since prostitutes are not allowed to operate openly in commercial parts of most cities in the western world, they quite often end up in residential areas and depend on pimps for protection. Just as prohibition of alcohol created crime, so does prohibition of the "victimless crime" of prostitution CAUSE crime, and often death. Criminal prosecution and prison terms have not stopped, and will never stop prostitution. Yet, law makers continue to criminalize a natural act between CONSENTING ADULTS.

While on the subject of sex between consenting adults: so-called "love-making" between a prostitute and her client has little to do with "Love." Only the natural sex-drive between two adults is acted out, to the financial benefit of one "partner," and to satisfy the natural sexual desires of the other.

A male human is attracted to the female body for the same reasons as a "courting" male bird when it is attracted to a female bird of the same species. Dictated by the same animal "instinct," human females get exited by the sight of certain male anatomical body parts. Young males in particular react likewise over the appearance of an attractive girl — this is certainly not "news." However, it is often misinterpreted as "love at first sight." Female classmates can be overheard talking about a handsome male's "buns" or conversation can be overheard when a male student is referred to as a "hunk," all part of so-called "girl talk." It is not "love" at all, with or without approval of religion, it is only the power of nature at work.

All of it is as natural as the physical attractions between male and female baboons, or for that matter, male and female whales or sea lions. However, it is interesting to note, only animals of the SAME species are attracted to each other. A sea elephant bull is attracted to a sea elephant cow, but certainly NOT to the sexiest human female, and visa versa! One possible theory for it is that over many thousands of generations, each species has developed another "instinct," making reproduction possible ONLY between the same species. Scientists have actually confirmed this phenomena. Still, it is sexual attraction for the purpose of reproduction, not love.

Starting in high school in their teens, most young women particularly in the western world, dress up and make up to appear attractive to their male counterparts, or possibly to compete with one another? Young men want to impress teenaged girls with their muscles and often get into fights over the most attractive girls, just as other mammals and some lower life-forms do. General appearance, a most attractively shaped body and the "chemistry" between a girl and a young man have the same natural effect on a human male, as sexual attraction does amongst other mammals. No mythical God brought them together, only circumstances.

Interesting to note here also, is the emphasis on stronger sexual attraction between YOUNGER people of opposite sexes. Or even when an older man feels sexually attracted towards a younger woman, the natural instinctive attraction is based on the greater potential for fertility of the younger woman. One would rarely find true SEXUAL attraction between a young man in his twenties and a lady in her seventies. In any case, all of it can be described as inherited primate instinct, and may or may not develop into a long lasting "love" relationship.

Young men tend to display physical strength and intelligence for the same natural reasons as all other mammals — to impress the opposite sex and to be a leader amongst their own. Men are attracted to the "sexiest" females in the same manner as other male mammals choose their most "attractive," (read healthiest) female partner for mating. These are the facts, and it has little to do with "Love." It is physical attraction for reproductive reasons only, to maintain the continuation of the species.

The point is: Contrary to biblical teachings, instinctive sexual activities between partners of ANY creature, is not "God"-given to humans only, and it may not be "Love" at all.

Of course, "LOVE" between human sexual partners DOES exist and really has a number of different origins. When sexual attraction between male and female humans develop into a more permanent monogynous relationship, (marriage) and it is resulting in a family, this kind of "Love" is no longer purely of a sexual nature. It is also as instinctive, as a bird family raising their young and defending them and their "mates" against intruders. It is as natural, and common as the lion family caring for their young. Or as the bird family taking loving care of their chicks until they are able to leave the nest to fend for themselves.

Most fellow creatures display "love" between parent and off-spring in one way or another. All primates and most other mammals behave in the same manner, including bats, sea horses and the tse-tse fly(!), along with millions of other non-human members of the animal world.

Humans often develop another type of "love" relationship, love between FRIENDS. It may be either between members of the opposite sex or of the same gender, and without "sexual" lust being a contributing factor. This type of "love" is by choice. We all had it happen to us, when we got along great from the first moment with new-found friends. At other times, a stranger could never get close to us. This type of friendship and "love" by choice has a similarity with attraction to a group by other "herd" animals. Or, when a single animal is rejected by the herd (his "friends"), and has to look for new ones to be "close" to. Love between human friends usually takes considerably longer to develop, compared to "Love" through sexual attraction.

Incidentally, homosexual "LOVE" is not an exclusively human behavior, other animals have been observed practicing it. We have only recently discovered that homosexuality is NOT a sin or a wilful choice of godless evil, it may be inherited through the DNA of the family tree, and therefore a natural phenomena. People with a genetic tendency for homosexuality should therefore be treated the same way as "straight" people, NO worse and NO better. In any case, what two CONSENTING ADULTS do about their private sex/love life should be of no concern to others.

The gentle touch of a pet by the human hand, better known as "petting" of a dog or cat, a killer whale or of a large carnivorous cat (in captivity!), will normally result in an affectionate response from the animal. Caring for our pets is one other form of "love," similar to the gentleness expressed by the normally carnivorous killer whales towards a human child, who affectionately pets this large "fellow mammal." The point of all this is to illustrate the similarities of emotions between ALL "higher" forms of animal life, may it be called "love" or simply, "affection."

It is said in the Bible that God "loves" all mankind. Considering the various real forms of "love" between "animal" species, including Homo Sapiens, "love" coming from a mythical, invisible and unproven superior life-form, is as difficult to accept as "love" coming from any "other" extraterrestrial visitor. The results of this hypothetical "love" are rather disappointing, when we consider all the suffering and misery through which many faithful believers had to struggle, alongside those who do not believe in God, or those who question His existence.

"HOMO SAPIENS" AND OTHER MAMMALS: IDENTICAL IN MANY WAYS, AN OFTEN OVERLOOKED REALITY

As we discovered earlier, mankind looks for a "leader" of the "herd," may it be a political or a spiritual leader — all part of natural animal behavior by most "herd"-animals, all the way down to some insect species. Unfortunately, most of our present SPIRITUAL leaders misinterpret and abuse the herd instinct, by preaching BLIND faith in some totally obsolete religious teachings and rules.

This practice led to torture and brutal punishment in the not-so-distant past of all those who dared to question religion, mostly during the medieval period in Europe. Fanatical religious leaders decided who was "evil," when a women (usually) was a witch or what form of punishment an "evil sinner" was to receive. Common sense and logic were usually not part of their vocabulary.

One can read in history books about the witch hunts

of Salem, Massachusetts during the late sixteen hundreds, when witchcraft was a capital offense. Scientific investigations have recently revealed that exposure to an LSD-type drug, consumed with the local water at the time in all likelihood has been responsible for the sometimes strange behavior of so-called "witches" — It is another example of innocent lives being taken, at the infamous "Gallow Hill." Once again — death thanks to blind religious faith.

In summing up, we are without question a member of the mammal family of animals, a usually carnivorous primate with more or less the same biological and evolutionary past, and the same present and future as our "sister"-species. I have tried to illustrate that the basic emotions and feelings are often identical, or at least similar, from parental instinct to fear and aggressiveness, from pain to sexual attraction. All animal bodies are subject to the same environmental and climatic changes, and humans are conceived and born in more or less the same manner. The lives of ALL mammals are primarily influenced by bacteria and other microorganisms within, and by environmental changes around us — NOT by a mythical "God."

Let's look at the facts one more time: Our bodies decay the same as all animal carcasses do, if left exposed to the elements. Many diseases are identical, or nearly so — dogs with arthritis, cats with cataracts, etc. As pointed out elsewhere, chimpanzees have 99% of the human genes.

So, what really DOES set us apart from the rest of the animal world? Partly by choice and necessity, our brains have been more developed, while at the same time other body parts have deteriorated, all as part of a natural, evolutionary process. Together with the brain, our hands have developed superior to other mammals', allowing us to use the most delicate tools. Our voices cover a much greater frequency range than those of most other creatures, while the rest of our bodies' functions continue to deteriorate due to lack of use. Sooner or later, when mankind may finally be entirely guided by fact-based logic — over religious belief — THEN and only THEN will the human brain be TRULY superior to the rest of the animal world!

Maybe genetic engineering will be allowed to make the necessary improvements to our bodies. Unfortunately, the readers and I may no longer be around when humanity finally comes to that conclusion. But then, who would have thought how fast millions could change their strong "belief" in Communism?

The human body will probably never completely overcome nature's imperfection, and the never ending war for survival between our antibodies and equally microscopic intruders. But science will continue to win battle after battle against diseases and aging, just as we have in the past. Scientists already understand one of the most important contributing factors to the aging process — an increasing break-down of the human immune system. As pointed out elsewhere, significant progress has been made. The human life-span has not only been extended, life for the elderly has been greatly improved. Without the help of scriptures.

Some scientists have recently come up with the theory that through elimination or decreasing the "deterioration of resistance" against infectious and other diseases with age, humans could reach a ripe old age of about 800 years. That is, provided humans stop killing each other during armed conflicts, through gun related fatal injuries and other "human error" accidents of all kinds. Other animal species already reach a much longer life span without trying. It is more noticeable, if the age of sexual maturity is applied as a reference to the overall life expectancy of most living creatures.

There may be some rare truth in the book of "Genesis," when the various ages of some biblical figures were stated as follows: Adam was to have lived to age 930, Seth died at 905 years of age, Cainan was 910, Methuselah 969 and Noah was 950 years old. It may be scientifically possible for humans to reach that age through a strange combination of herbs, and even poisons, which were to have pro-longed life in the biblical days. However, if human age in Genesis was given in actual calendar years, the "real" age of our planet Earth was surely given most incorrectly.

As I indicated earlier, I accept the theoretical formula to calculate the true age of an animal by multiplying the age of sexual maturity or puberty, to arrive at a potential life span. Adding a few more examples at this time: a female gorilla reaches sexual maturity at six years of age and has a life span of thirty calendar years. Compared to sexual maturity of humans at about twelve years, the formula therefore is simple: at one-half the age of sexual maturity, thirty calendar years become sixty chimpanzee "years." According to that, humans should have no trouble reaching the ripe old age of one hundred and twenty calendar years! Taking another mammal at random, the giant anteater is sexually mature at two and one-half years of age. Their average life span is twenty-six calendar years, but compared to chimpanzee "years" they would live about two hundred and fifty anteater years, since their calendar age is almost the same as a chimpanzee's, but less than half the age for sexual maturity. Other formulas may give different results, but most mammals live a longer and healthier life, no matter what calculation is used.

Getting away from mammals, certain large land turtles reach an age of several hundred years and all that without a balanced diet and supplementary vitamins! As a matter of fact, most turtles can live without food for weeks, showing no ill effects. It is remarkable that ALL living creatures age and eventually die due to decreased lack of resistance against infection, or inability to spot and to out-run predators! Although some animals age faster than Homo sapiens, but when an acceptable formula is applied most of them have a relatively much longer life span under very adverse conditions — their bodies are not as fragile as ours.

For this reason alone, the biblical quote giving "man

the domain over all creatures" bears no logic. Believing it caused extinction or near-extinction of countless animal species, right to this day. Mankind needlessly killed millions of whales, until it was almost too late for some species of this intelligent mammal. Our ancestors killed millions of Bisons senselessly and caused them to become an endangered species at one time. We don't have to go back into history at all — overfishing during recent decades caused a crisis at the east coast. Natural evolution can only compensate to a point, while man-made or natural disasters CAN cause the extermination or near-extinction of any number of the hardiest of animal species.

Only through religious teachings and traditions, mankind "believes" to have the right to kill other creatures at random, and NOT just out of necessity as a food source only. Modern man does NOT have one logical reason to "balance" the wildlife through hunting — nature took care of that for over 300,000,000 years. And long before man appeared with religious "wisdom" on this planet.

Native people of Northern Canada and the South American jungles have been hunting over the centuries for necessary food and clothing ONLY, and none of the hunted animal species have ever become extinct as a result of it. On the other hand, a few hundred thousand religious church-going settlers in the United States and Canada senselessly killed about 60,000,000 Bisons in a matter of decades!

Fortunately, only in recent years intelligence and logic have taken over in some nations, when it comes to killing of wild animals for fur coats, hats, ivory and horns. Various governments have started to prohibit the killing of various wild animals, such as elephants, rhinos and cheetahs. Yet, poachers still dangerously decimate their numbers for profit only, through the sale of fur and ivory to "civilized" countries where ancient traditions still dominate over common sense.

Mankind's killing and breeding of animals for food is a necessary evil of nature, just as other carnivorous animal species kill and sometimes even "breed" other creatures for food. But that is where the laws of nature end, there is no logical reason for the mass-slaughter of animals for any OTHER purpose, including "sport-hunting" — unless we want to use illogical biblical teachings as "reason." Since we are able to manufacture the finest artificial furs and "artificial" ivory, we certainly do not "need" body parts from whales, elephants and seals. Often still near the end of the twentieth century, needless hunting of animals as a sport is done by very religious individuals, sometimes before or after prayers.

Additional evidence of the close relationship between Homo Sapiens and other mammals came up when medical doctors successfully transplanted baboons' hearts to dying heart patients and with limited success. Only a few decades ago this experiment would have been considered evil by firm religious believers.

The human life span is given by one version of the Bible at about 70 years. As detailed earlier, there is absolutely no reason why we could not achieve the same life span as some of our relatives in the animal world. We already have more living 100 year old people (centurions) on Earth now, than we ever had before, over the entire recorded human history.

In consideration of the latest scientific findings, one question comes up: does old age cause deterioration of body resistance, or does deterioration of body resistance cause old age? Or are other factors involved, such as excessive solar or other harmful radiation from intergalactic sources, the so-called "Cosmic Rays?" Maybe, as the genes are copied from generation to generation, a certain deterioration occurs leaving the new genes weaker than the originals? Unfortunately, we may never know, since Earth will never have a "perfect" environment, WITHOUT bombardment from gamma rays, etc. Unfavorable conditions often outweigh conditions favorable for life.

By applying religion too often and without fact-based logic, mankind will only slow down the search for discoveries to better our general health and to increase the life span. Only modern science will continue to find ways to help nature to catch up with degenerative evolutionary changes. To this day, some of them are caused by lack of knowledge or through man-made environmental pollution. Without doubt, we will continue to learn and discover the facts of nature and the universe, without interference from religious teachings and prophecies of doom and gloom.

Research will continue to run into stumbling blocks along the way, scientists will learn the "hard way" more than once, just as a child does in the first grade. Our scientific knowledge is really still in the "first grade," considering that most major discoveries were made only in the past 50 to 100 years. Merely a few hours in human history, compared to 5000 to 10,000 years of recorded history. Recently however, the total sum of human scientific knowledge has doubled every two and a half years.

As humanity stumbles along, medical science has discovered lately how to successfully cure and treat diseases WITHOUT drugs, by using "encapsulated cells." A combination of living cells and "compatible" synthetic materials, "biohybrid" capsules may be one of the most significant medical advances of recent years. Some disorders such as diabetes, are up to the present still cured with pharmaceuticals or organ transplants, both resulting in various side effects. Biohybrid capsules appear not to cause any short or long-term side effects to the rest of the body, contrary to drugs and organ transplants. Biohybrids have no supply problem. Only a few cells are needed, and are often derived from "other" animals. The capsule physically protects the cells from attack by the immune system, and suppressant drugs with possible degenerative side effects are therefore not required.

To pick an example: diabetic patients of the future, after a one inch, three ounce biohybrid capsule is implanted, will no longer require daily insulin injections. This "artificial pancreas" contains a coiled semi-permeable membrane tube. Transplanted pancreatic cells live on one side of the membrane, producing the much needed insulin. It travels across the membrane to where the blood circulates.

The blood continues to provide the pancreatic cells with oxygen and glucose, but immune cells are too large to penetrate the membrane. The system is self-regulating, since the cells respond to sugar levels appropriately. The artificial pancreas is currently undergoing review by the F.D.A., but the developers should have little or no problem, since no chemicals or artificial ingredients are involved. Another promising new method of pancreas replacement is the implant of healthy pancreas cells into the stomach. We have future medical science working for us today, but this type of news does not appear to make headlines.

The moral of this story is to confirm again that the human body, not unlike other animals, is virtually completely under control of micro organisms, beneficial as well as harmful species, with a continuing battle between them from conception until after death. All mammals and other life forms are effected the same way, the almighty one-cell organism controls all lives. No scripture teaches us these important facts, medical science had to discover them. The words "bacteria," "DNA," "Genes," "Mutation," and "(harmful) radiation" are not in the vocabulary of the holy Books. Prayer does not produce insulin, it does not prolong a diabetic's life. Scientific research alone will continue saving thousands of lives daily.

PREDICTABILITY OF DISASTERS THROUGH SCIENCE

Californians are prepared for major earthquakes through scientific studies and by historic events. One of them will be extremely powerful and devastating, between today and about thirty years from now. At the same time, scientists are on the lookout for stray asteroids or large comets which could have disastrous effects on all forms of life, in the event of a collision or near-collision with our planet. About 65 million years ago, mankind's evolutionary predecessors did not have this knowledge. As a result of it, large dinosaurs and many plants and other animal species became extinct, after the most drastic climatic changes to this planet.

Perhaps the upheaval was caused by Earth's collision with an asteroid or large comet. At least the iridium deposits found in an almost world-wide "dust" layer of what is presumed to be of extraterrestrial origin, dating back approximately 63 - 65 million years, seem to be strong evidence in that direction. It was a long time before the so-called mythical "biblical flood." More about it later in this book.

Only the hardiest and the smaller land-mammals survived, along with most sea creatures. Recently, during an expedition through the Gobi desert in Mongolia, archaeologists found eighty million year old remains of shrew-like small mammals, AMONGST petrified dinosaur bones of the same age. Such is the latest evidence of the existence of mammals at least 80 million years ago and further proof that these small creatures actually survived until modern days in form of shrews, mice and rats. Yet, mice, rats and hamsters are so closely related to homo sapiens today that most life-saving medical research is based on initial experiments with these animals in the laboratory. Were our ancestors 80 million years ago amongst the "shrews" of that period?

After much controversy between scientists, there is now growing consensus that Earth was indeed impacted by a six-mile-wide asteroid (alternatively, but less likely, a comet) about 65,000,000 years ago. According to evidence obtained, the resulting explosion must have caused the Earth's climate to change drastically for a number of months or years. Today, earth-bound scientists can not only forecast an impending asteroid impact, they are also able to deflect the object by slightly altering its orbit, through a massive "friendly" nuclear explosion. A sort of modern-day "Noah" with real scientific knowledge! Unfortunately, we cannot find the instructions for prevention of this hypothetical, but highly probable disaster in any holy book.

Also predictable would be the return of an ice age, which earlier caused the extinction of a number of entire species possibly as late as 25,000 years ago. At that time, there was nobody around to predict it and to evacuate some of the creatures. Today, we have the capability to prepare for at least some climatic changes of this magnitude and to save humanity and other species in the process. The slightest change in our planet's position, in relationship to the sun could cause yet another ice age in one hemisphere, resulting in pro-longed unbearably hot desert conditions in the other. Some species would again survive without our help, others will not. After all, we already have plant and animal life surviving and thriving in the most hostile areas of the world. Quite a number of life-forms on land and under water had survived the "big one" about 65,000,000 years ago.

Recent studies have revealed, that the "wobble" in the Earth's rotation may increase to dangerous proportions and may also result in fundamental changes in weather patterns. Possible causes could be change in gravitational forces from a crashing, or "fly-by" extraterrestrial object such as a comet, an asteroid or a very large meteor, causing the slightest alteration to the existing wobble. If the change is significant enough, an increase in the glacier coverage on either the antarctic or arctic regions would occur over a number of centuries or possibly over a period of a few decades only.

If the increase in the weight of the ice-caps at either polar cap becomes enough compared with the opposite polar region, the wobble increases accordingly. This reaction is comparable to an out-of-balance automobile tire and it could result in a dramatic increase to existing climatic extremes, to the point where many life-forms will either become extinct or will have to migrate to more moderate areas of the planet. The birds are thought by some to be ancestors of dinosaurs and they have the edge on this one, they can fly to a moderate climate. The capability to fly is quite probably why these small descendants of flying dinosaurs survived so far.

Fortunately for mankind and most other animal relatives, scientist can now predict with reasonable accuracy, if and when an impending natural disaster of this type and dimension will be triggered. In contrast, biblical prophecies have warned of certain natural catastrophic events only as a result of "God's anger." The dates of biblical prophecies are very vague and can be translated in any way a believer wants to.

Scientists may not always agree on various theories today and in the future. However, without application of religious teachings, we are now able to forecast all major changes to our Earth's environment. We were unable to do so only thirty to forty years ago, at a time, when the present environmental mess was created. Scientists are currently rushing to correct the results of the old mistakes from the "carefree" days of the fifties in the western world. Thanks to increasing scientific knowledge in the eighties and nineties, and without help from religion, mankind is gaining ground slowly. As mentioned before, science is not perfect, yet it is doing much more for humanity than religious teachings ever have and will.

Humanity has another, less scientific source of predictions. Not necessarily religious people, but simple everyday individuals who seem to develop the capability to foresee serious and sometimes cataclysmic events to our planet. Many of them, possibly most, cannot be taken serious, as a researcher by the name of James Randi has been able to prove through years of investigations. To start with, two thirds of the predictions were found to be generalities, which will happen or are very likely to happen somewhere along the line anywhere and to anybody.

As sure as the sun will set every evening, we all will have poor health at one time or another, nearly everybody has financial difficulties at some point in life or have problems with family members or friends at one time or another. All of us have had, or will have some sort of accident. All these and other events are occurrences which everybody could predict.

Therefore, I will only concentrate on the two most creditable individuals, the psychics Edgar Cayce and Gordon-Michael Scallion. Before Ed Cayce died in 1945, he had made a number of remarkably accurate prophecies, too accurate to be ignored. Doctors tried to prove him to be a fraud, but were unsuccessful. In his days, he foresaw the 1929 stock market crash about a year before it happened. One time he saved his own life, when he foresaw imminent disaster just before he was ready to enter an elevator. Seconds later, the cable snapped and all occupants were killed.

During over 14,000 "readings" he foresaw with remarkable accuracy many minor and major events in human history, such as the approximate timing of the collapse of communism in Russia (yet, he correctly made no mention of communism in China), and many others. His most significant capability was "distance" healing of apparently incurable patients. All he needed, was the patient's name and address and he was able to pin-point the source of the ailment in most cases, and usually when conventional doctors failed to diagnose the problem. He performed this practise only very hesitantly, since he could not really understand his own powers. He really seemed to be afraid that his diagnosis could be wrong and harmful to the patient. Also, he had to go into a trance-like, very exhausting "sleep" to be successful.

I chose his example only because he also foresaw one major catastrophe about fifty years before it is to happen — two nearly simultaneous great earthquakes at the end of this century. One, of course, is the scientifically expected massive earthquake in California and the other, also predictable, a similar quake for Japan. Scientist agree on the approximate timing, particularly for the California quake.

He foresaw the destruction of Los Angeles, San Francisco and strangely enough, New York City and large areas along the entire U.S. east coast. He prepared a map of the Unites States after the great cataclysm at the end of the century. Considering the accuracy of many of his earlier predictions, one should not take this prophecy all too lightly. Particularly not, when scientists also predict the same event (the big California quake).

Even more remarkable are the prophecies from another well-known psychic, Gordon-Michael Scallion. He too foresees the great California quake and major shifts in the earth's crust, also near the end of the century. He acquired his unexplainable psychic powers in 1979 and has been right in the past with the majority of predictions of minor and major events. He published his prophecies in 1991 with another map of the United States, how they appear AFTER the cataclysmic disaster near the turn of the century. According to his forecast and the map, and similar to Cayce's prediction, a major portion of the western U.S. will disappear with part of the Rocky Mountains emerging as the new western coastline, and the Pacific reaching as far as Colorado.

His prediction includes a shift of the earth's crust of such a magnitude, that the Great Lakes will merge to the same level and will partly drain through the Mississippi Valley into the Gulf of Mexico. Looking at his map, large parts of the East coast will be submerged, all the way from Maine to Florida. The northern part of up-state New York will be submerged, while the adjoining parts of Ontario, Canada appear to be spared. The map leaves the impression that the land masses east of the Niagara River will sink hundreds of feet, while land west of the river will remain unchanged. Of course, this event would in all probability cause the disappearance of Niagara Falls.

Again, I have picked the prophecies of the two psychics only because they partially agree with scientific studies and findings. As far as the changes to the eastern U.S. are concerned, there is no scientific back-up for this change, although it is remotely possible. It is up to the reader to leave some of it up to "belief" or to accept it as "fact." Any optimist, such as myself would probably hope, that this prophecy is one of a few which will NOT materialize. The fact is, scientists have no indication that major shifts in the earth's crust will occur, other than along the west coast and possibly in Japan through massive earthquakes.

The other prediction by Gordon-Michael Scallion about a possible shift of the Polar regions from north to south and visa versa, has to the best of my knowledge no scientific basis. Only a collision, or near-collision with a massive asteroid or a large comet could possibly cause an instant reversal of the polar regions. Yet, scientists to date have no knowledge of a pending collision around the end of the century. However, he has been right with the predictions of the eruption of Mt. Pinatubo and the 1992 California earthquake.

A few lesser known individuals seem to have the unexplainable power to foresee coming disaster also AND they can prove its accuracy. As I said earlier, there are a number of fraudulent psychics, and investigation usually disproves the validity of their "visions." Not so with the following example of one of the few true spiritualists: David Booth had reoccurring dreams about an air crash. He saw the type of plane, he recognized it to be an American Airlines jet and he saw it in an unusual position just before crashing, with one wing straight up and the other straight down. After the nightmare repeatedly disturbed his sleep he became quite concerned and contacted Paul Williams of the F.A.A. by telephone.

Unfortunately, the location of the crash was not clear from his "dreams" and nothing was done about it. Only days later he could see the shocking TV news bulletin about the crash of an American Airlines jet over Chicago with the exact same position of the wings during the last seconds, as he saw it in his dreams. The fact that he had reported to the F.A.A. before it happened is proof enough that his vision was indeed not fabricated.

And finally, the great prophet Nostradamus also foresees disastrous earthquakes and volcanic eruptions for the turn of the millennium, if the interpretations of his 500 year old four-line prophetic verses can be considered accurate. He too refers to a cataclysmic series of events near the end of this century and he blames the "great conjunction" of the planets as the cause. A planetary conjunction occurs, when two or more planets are all lined up in one row, causing interplanetary gravitational forces to act upon each other. Earth has been in this position before, and the effects were negligible, except higher than usual tides. Nostradamus of course did not know how far the other planets really are from Earth and how small the gravitational pull from other planets during a conjunction really is.

Today we know that unchecked human technology has been, and still is often causing damage to our environment and to our health. Quite often mankind is still playing with acts of environmental self-destruction and it can not always be blamed on LACK of knowledge, such as pollution of lakes and oceans, clear-cutting of northern and rain forests, over-fishing of cod, etc. Scientists discovered only a few years ago how much damage we really did to the Earth's ozone layer, while we were using fluorocarbon gases for aerosol cans and refrigerators. However, much is being done to correct past mistakes and to avoid future damage to the environment. And all the damage done so far by mankind would not equal the massive changes to our planet, as foreseen by both psychics. Even a nuclear war would not cause such drastic changes. Only sudden, often unexpected sunspot activity results in more damage to our atmosphere than all other causes combined. To be exact, the sun causes over 99% of atmospheric changes.

To sum it up: No superior "God" has warned us about any particular recent natural disaster — the victims and scientists found out the hard way, and they will find out again in the future. In any case, biblical prophecies refer almost exclusively to the relatively small area of the biblical Middle East. Maybe between a few "true" psychics AND modern science, mankind will come up with the right predictions and solutions for North America, South America and the rest of the world?

Another well recorded fact is how often we have been subjected to weather extremes, volcanoes and earthquakes in the past. Earth has been suffering many times under massive natural disasters, long before man appeared. Earth survived single disasters millions of years ago, and of greater magnitude than those of the past 3000 years combined, yet the planet is still here and so are countless life-forms. No matter what, some life will always survive, including some humans, (except, when the sun becomes a "red giant" in four to five billion years from now). This includes what could be called the "Armageddon" of 65 or so million years ago, which came and went without biblical or psychic predictions.

Once again, the reader has to choose between FACTS and BELIEF; those who prefer to accept as fact the prophecy of "Armageddon" in five or six years from now, should prepare by possibly purchasing ex-Navy submarines and fill them with supplies. The only reasonably safe place for survival of this hypothetical event IS under water, (aside maybe from being on top of a mountain) away from all areas, which could be effected.

Those who only "BELIEVE" in it, whether it is based on RELIGIOUS or PSYCHIC prophecies, should do nothing (as usual) and could leave it all to "God's will." And finally those who just BELIEVE in the predictions of the psychics, may want to be at least prepared and watch for possible preceding changes to the planet, listen to scientists and watch for possible, as yet undiscovered dangerous asteroid movements. I consider myself as falling somewhat into THIS category, but I am not yet planning to build a submarine!

WHAT IS THE "SPIRIT" — WHAT DO WE KNOW ABOUT IT?

There is no doubt about what happens to our physical body, after natural or accidental death. All atomic and molecular matter from our bodies remains intact and unchanged. That includes calcium and other atomic matter. Almost the entire 75 - 80% water content of our bodies, (about 58 liters for an average adult male), evaporates and mixes with moisture in the air, with humidity, and with fog; unless a corpse is immediately buried or otherwise prepared to prevent evaporation of the water content. Either way, most of the body's moisture will end up in the clouds and later in lakes, oceans and rivers, and it will be coming down again somewhere as rain or snow.

This cycle will repeat itself all over and some of a corpse's original water molecules may be part of a newly forming human baby, via intake of liquids by the expecting mother. "Matter" never gets lost in nature. Other decaying, non-liquid "mineral" matter will eventually supply a nearby tree or a deep rooted plant with nourishment. Maybe it will take centuries before all matter from our bodies is "recycled," but it will happen. Much in the same way as any other animal corpse decays.

If a body is CREMATED, the process is accelerated immensely, almost the entire moisture from the corpse will evaporate during cremation. Death in a lake or ocean will similarly cause the body's water to be released directly and it will be mixed with the surrounding body of water. Immediate burial of a body will have the opposite effect, the body's water contents will be trapped underground for many years and will only be released very slowly. The process of decay will be left almost exclusively to still active bacteria and other microorganisms in the body in this case, and not as much to the "worms" as one might be conditioned to believe. However, wherever the remains of a body eventually end up, it is all part of a natural cycle.

Nobody, including scientists, has as yet been able to find definite proof of where our so-called "spirit" (or soul), goes AFTER death. Equally, science has not been able to prove its existence, during life. It is not a physical entity as the rest of all body "organs." Most religions refer to the human soul with hypothetical predictions of what happens to it after death. The fact is, humans do have certain "spiritual" powers throughout LIFE, some more than others, but it is not clear WHAT they are. We cannot see the spirit, we cannot readily measure it with measuring devices — unless we want to refer to special photography of an "aura," which is only visible to a few psychics.

Scientists have developed certain theories about the existence of a soul, but they have not been able to prove any of them. The closest we have come to the possible "location" of our soul is in the brain. During our life our "soul" is guided by our senses: interaction between sight, sound, taste, smell, touch and also by memory. We also DO know, there is no memory and no feeling from the time BEFORE BIRTH. Our body and the brain started out with "re-cycled" atomic matter, a great proportion of water molecules, before conception.

Later, two microscopic cell structures, an egg cell and a sperm-cell formed a new life and the genes were transferred. Primarily the waters of the oceans, lakes and rivers made the creation of life possible, together with minerals and chemicals, all of which had been "absorbed" by the parents in one way or another from food and drink. The human embryo continued to absorb re-cycled water, calcium, sodium and all other minerals, through the mother's body, until birth.

At what point we acquire a "spirit" (or "soul") is not clear from a scientific point of view. Is it during conception, as an embryo, as a fetus or as a new-born? We have numerous theories, religious AND scientific, but we have no proof one way or another.

One of the most logical theories on the subject is the following: the "spirit" of all higher life forms grows with the entire body and the brain and it also dies and decays with the body. Of course, if we accept this rather logical theory, the "soul" must end with the body (or brain) functions, it will be non-existent after death, as it was before birth (or conception).

We need the ENTIRE brain to function, to re-call memory, to see and to feel, to hear and to taste, to make decisions and to interpret what our nerve and brain cells tell us. Only during a healthy, normal life can ALL of the body's cell structures function. After biological death, with the vital nerve and brain cells dead, along with the rest of the body, a living creature can no longer feel, see or hear, think or reason. The only really unknown quantity in this condition is the MEMORY. Therefore, is the "memory" the "soul," or the spirit of a higher life-form? Possible, but a long way from being proven fact. If one wants to belief in this theory, the question comes up: what happens to the memory/soul after death, is it transferred with atomic matter to another person like memory chips are transferred from one computer to another? The answers can not to be found in religious books.

Unfortunately for religious believers, and to add further logic to this more scientific theory, we have no fact-based reports of "returning" souls (memories). Nobody has come back from *biological* death and was able to tell us about what happened to them during "after-life," and giving us proof of their existence after death. One can rely on religious theory and teachings after careful consideration of all known facts, but the believer is really walking on very treacherous ground. In summing up — the human body and those of other primates ceases to function after biological death, the nerve cells die along with the rest of them. And nerve cells are responsible for an animal's senses.

To scientists' best knowledge, our brain only starts to function slowly part-way through the development of the embryo, what becomes after three months the "fetus." The fetus starts to react early to touch and other certain primitive external events, such as loud noise or pain during an accident by the mother. This behavior is strictly instinctive from an "un-trained" brain and it will continue on for the rest of our lives. All mammals and their fetuses are more or less reacting in the same way. These are established FACTS. What we do not know, is WHEN our brain is enough developed to go beyond instinctive reactions. When do we "know" what we see, when do we start to memorize? At what age can we start to use logic? How can we use "logic" through our soul AFTER death, if one believes in "life after death?"

Maybe the answer can be found, when we stop to think that homo sapiens and other primates develop from two single cells at conception, containing all genetic information. Over a relatively short period of nine months, a human embryo and fetus develops from two cells into hundreds of millions of body cells. That includes brain cells with their increasing capability to function, as the numbers increase. After birth, the brain cells, along with other body cells multiply much less rapidly, and the human mind starts to accept increased "training" from the surrounding sights and sounds and from the parents.

Reasoning and the senses also develop at a slowing pace along with the body, so why not the "soul" of the individual? It is quite apparent that only a very fine line exists between "reasoning" and "instinct" and it is often difficult to tell the difference between the two brain functions. Our understanding of the existence of reasoning by all higher life forms is becoming more evident lately. The old religiously inspired myth that only mankind can "reason" and other animal act by instinct alone has long fallen into obsolescence.

Using logic and facts again, the body parts AND the brain appear not to start with a "Big Bang" at the time of birth or at conception. It may be some time after birth, when a child begins to realize the capability to make decision after seeing or hearing certain events. It seems therefore that the so-called spirit also evolves SLOWLY in the same way, as a body grows from two single cells. After all, there are few reports of sightings of fetus or baby "ghosts," if these "sightings" can be substantiated at all.

This still leaves the question: where does the spirit come from in the first place and exactly when do we receive it? To our knowledge, it is probably not transferable like a REAL body part. Medical research may find out, if and when we are capable of brain transplants, will the "spirit" be transferred with the brain?

That brings me to what many people swear to and accept as fact: RE-INCARNATION — many are convinced, they have been on this Earth before. Under hypnosis, individuals have described distant locations and countries and spoke languages, which they could not possibly have known from their "current" life. Facts maybe, but as yet unexplainable.

Let us back-track. What IS the human "Spirit?" As stated, it is not visible, nor is it a measurable scientific fact. The strongest electron microscope cannot "see" the human spirit. Yet, we DO know, an invisible power exists, which is "You" or which makes me "Me." A controlling power which makes only ME feel pain, when it is inflicted on MY body. The power, which controls MY emotions and which causes an individual person or animal to feel hunger, to feel tired, to feel sexually aroused, to cause voluntary movements. Also, the power to interpret sight, sound and touch, including some semi-instinctive, in-voluntary reactions including sleep, digestion, hunger and sex drive.

Other "mind powers" are completely under our own control and can be shut off or switched on through a strong will. Or through a strong "Spirit?" There seems to be no clearly defined border between instinct, emotions, will-power and spirit, unless one considers all of the above as part of the "Soul" or the "Spirit?" During our lifetime, the "spirit" controls good and bad emotional action, such as drug addiction, smoking, drinking, over-eating and certain other strong desires. For example, near "fanatic" search for wisdom, through what could be called an "excessive" learning drive, or a housewife's desire to engage in "over-kill" house cleaning, all can be caused by seemingly uncontrollable urges or emotions. Along the same lines, another spiritual emotion is the acceptance or rejection of a religion, other than simply taking up the same faith as the parents. And finally, the spirit can also cause criminal and unscrupulous "emotions."

No matter how hard it has been tried, the existence of a spirit AFTER DEATH is nearly impossible to prove. To me, the "spirit" of a person continuous to exist only in the memory of close SURVIVORS, the family and friends. A murder victim will be remembered a long time after the tragic death and by many. In contrast, an executed killer will NOT be remembered in a pleasant manner and by possibly only a few close relatives, and then with mixed emotions.

While on the subject of murder and as example of a negative function of a "spirit": A killer planning premeditated murder knows exactly that murder can cost him his own life, depending on which country or state he lives in. His mind becomes so uncontrollable that in the end a murder is really committed anyway without consideration of the consequences. An example of how the human mind ("spirit") can be very strong in either a negative or a positive manner. In a positive way, so-called "willpower" has been known to bring cancer into remission, amongst other changes to physical and mental health.

Let us look at the brain at this time. It is a combination of cell "colonies" with very precise individual functions. One such colony are the "neu rons," some figures given are as high as 100 billion of them. Without listing all other brain cells, the neurons' function is to interact with certain chemicals and minerals. Two of them stand out in my mind: "Dolfemin" and the hormone Adrenalin, of which surges in interaction with neurons cause humans (and presumably all primates) to "feel good" or to "fall in love," along with the creation of other emotions, good or bad. Our entire character is shaped by this brain chemistry and the inherited genes. It may also be what we call the "spirit," at least in a living person.

Medical researchers in Hanover, Germany are in the final stages of applying — experimentally — the "Neurobionic" theory. It is another form of genetic engineering, when certain micro chips are implanted into the brain, thereby restoring a dysfunctional nervous system. In other words, totally disabled patients may be able to use their limbs again, while they currently have no chance of recovery, no matter how much therapy is applied. In the words of one of the developers of this process, Hans Werner Bothe, this discovery may be as important as the development of nuclear power was. He warns however, this process could be abused by manipulating brain functions of individuals by unscrupulous criminals, just as nuclear power can be abused.

Although much has to be discovered yet, science understands the BASIC brain functions, the interaction of brain cells, chemicals, hormones, oxygen and many other ingredients, it is the only possible "evidence" of a "spirit" in a human being. If one of the links in this incredible chain dysfunctions all the rest of the "components" usually break down also. Biological death is the final seizure of all brain functions and consequently the "spirit" of an individual. All primates' bodies AND brains have developed this way over hundreds of millions of years through evolution and the process is still continuing until the end of the planet and of all life along with it in a "mere" five billion years. No evidence of a mythical God can be found as a guiding factor for brain function, no more than external "demons" or "evil spirits" are responsible.

If the human "spirit" would have to be described by me in the least scientific terms, I could only do it in the following manner: an invisible central "computer chip" in the brain, made up from selected atomic matter and capable of emitting and storing electrical impulses, sometimes unpredictable and irrational. It controls all senses, the memory, and of course all emotions. But no matter how I look at it, there is no evidence of "spiritual" life after death, no "evidence" that the spirit of an individual continues to live on AFTER death. At least not in a hypothetical "heaven" or "hell," or on Earth. However, some may accept as "fact" the alleged power of certain individuals to contact the so-called spirits of some of the dead.

One possible theory for the continuation of a "soul" after death is a form of electrical impulses emerging with atomic matter from a deceased body. Possibly coming from the water molecules or as part of atoms and their electrons — at least, in our mind, we have some "physical" forms to deal with. As mentioned, atomic and molecular matter from all living bodies remains on the planet, unless enough heat is created somewhere along the line for a nuclear reaction, thereby transferring certain elements from one to another or converting them into energy. More on that later. Considering it all, is it reasonable to assume therefore, if the soul is BELIEVED to exist after death, and if it is transferred to another living creature, why not consider the only factual way possible — through atomic or molecular matter?

I have considered this possible solution, since it is difficult to ignore some examples of reported psychic contacts with the dead: George Anderson of New Jersey is one of many psychics, who claim to have the unexplainable power to "see" and to contact the dead. His powers, and that of many others, can not be written off completely as a hoax. To watch him and others on Television or to read about him, should be convincing enough for many to give some creditability to his powers and to the possibility of spiritual existence after death. Some of the detailed "visions" he and his colleagues are experiencing during contact with the "spirits" of total strangers, are usually impossible to fake. However, the largest percentage of "contacts" by many others ARE either intentionally faked or imagined, still leaving us with a number of unexplainable contacts with the deceased.

Another alternative has been suggested whereby the spirit of a person is transferred to a newborn baby through

DNA, or via other genetic material. However, this theory has a serious flaw: Hypnotic psychics could only contact those spirits, who are closely related to the subject under hypnosis. Based on this theory, spirits can only be transferred from one of the parents, together with all the other genes which determine the physical features, health and mental capacity. In reality, and under hypnosis, subjects have been brought into contact with other relatives and near-strangers, more often than with a deceased parent or grandparent.

One possible way to explain psychic "visions" of the dead, would be the process of "evaporation" of a body taking place after death. The escaping water molecules may possibly contain some "spiritual" information within the hydrogen or the oxygen atoms. An unknown, and therefore hypothetical type of radiation from this atomic matter may under certain conditions temporarily re-assemble into the shapes of a deceased individual.

Of course, speculation is wide open, since no scientific or religious explanation exists. Also, the so-called powers of "spirits" of the dead (or "ghosts") are very limited and usually temporary. Based on the reports from psychics, not everybody can be contacted, leaving the question: what happens to most of the (currently 5.5 billion) spirits, after the "owner" has passed on? The world would be awful crowded which ghosts, if EVERYBODY was to remain here in form of a spirit after death.

If we accept as fact the existence of a human "spirit" (during life *and* after death), another question comes up: Do all animals, particularly our closest relatives, the rest of the primates have spirits? Everybody has seen at one time or another a fast decaying dead mammal at the side of the road in hot weather. The same process takes over a fatal human accident victim, if the body is not discovered soon after the tragedy. That brings up a new question, particularly for firm believers in "life after death": If VISIBLE body parts from ALL mammals react in the same manner during life *and* after death, why should the "soul" behave any different?

All mammals including humans have nerve cells, their brains translate the messages received from their eyes, ears and noses. The question is: why do so many humans categorize other mammals as "they are *only* animals," without a "soul" or a "spirit"? If we than accept the conclusion that all animals not only have a soul during their life time, but also AFTER death, it could get a bit scary to think that the ghosts of all slaughtered cattle, pigs and chicken could be haunting humans, their killers?

As mentioned, the 75% or so water content of a mammal's dead body evaporates very quickly without a protective cover and if left exposed to the elements. As the evaporated water is no longer replaced through food and water intake, the remains dry out quickly, including the bacteria which help to speed up the decaying process. As stated earlier, the water molecules released from the corpse will after evaporation come down again as rain or snow.

The cycle is re-started when the same water molecules, absorbed by plants or other animals (and humans) through the food chain, evaporate again. Most of the (rain) water will remain in lakes and oceans for a long time, considering the large area of our planet covered by these bodies of water.

Coming back to an earlier conclusion: since most animals live and die in the same way as homo sapiens do, (evaporation of the body fluids), ALL living (higher?) life-forms must have some sort of spirit? Each individual animal feels pain, hunger, sexual urges, etc., just as a human being does. It all happens through the same invisible "personal computer chip", which controls the functions of each individual creature. An animal cannot feel pain inflicted on another, no more than homo sapiens can. To repeat the process again: all mammals are conceived and born in the same manner as humans, all of them live and die the same way AND the bodies decay in the same way — so what happens to the other mammals' "spirits" after death?

The question again is for firm believers in an "after-life": If everything else is more or less the same, why should not ALL higher life forms have a "spirit" during life and after death? It appears, the holy books conditioned believers on human "after-life" only. The "Central Computer" which controls their emotions and instincts, makes the pet dog feel only HIS pain, HIS hunger, HIS anger, HIS sexual urges, etc., just as his human master feels only his own emotions, pains and urges.

With the similarities between Homo Sapiens and other mammals as remarkable as they are, why should the possession of a "spirit" be the only difference between man and animal, when everything else is the same? Because it was "written" 3000 years ago in the Bible? Here is another interesting and notable similarity between humans and mammals: all their bodies contains an almost identical percentage of sodium (salt), and as much as that of our oceans. Also, a human embryo in its earliest stages of development briefly displays gills and until the moment of birth, the fetus does not breathe air, just as most other mammals do not and really cannot breathe air while in the mother's womb.

When we refer to the spirit of humans and other mammals, the biblical phrase "ashes to ashes" somehow came to my mind. Maybe it should have been at least more accurately phrased as, "water to water," considering the much greater water content of our body, compared to "ashes" (minerals and metals). The water content of any living creature is more quickly recycled than "ashes," during life AND immediately after death. The much smaller mineral content of a human cadaver is not only re-cycled much slower, if at all, it may not leave the remains at the burial site for centuries. Much of it will be left in the skeleton or ashes. Therefore, the soul of a deceased "mammal" could only be transferred through the "water cycle," if at all.

Given the proper environment at burial, well preserved human "bone dry" skeletons have frequently been found, hundreds of thousands of years old. Thanks to the LACK of water. When did THEIR spirit leave their body — with the evaporating water? Or is it still within the skeletons?

Interestingly, the frozen body of a young male has recently been discovered in mountain glaciers. The individual was thought to be a victim of a recent crime, until archaeologist established the time of his death at about 5,000 years ago. He had with him a number of simple tools and wore man-made clothes and boots. This discovery brings us back to the unanswered question about the existence of a spirit AFTER death: Is it an atomic structure, just like the rest of the body? Is it of unknown chemical composition, allegedly visible as a "haze" to some, but invisible to most? In this case, did the after-death "spirit" stay with the preserved body, with the visible body-parts?

Psychics frequently describe spirits as a nebulous haze in the shape of a person. If so, does the spirit remain with the deceased body for a short period or for a long time, or forever, just as other molecular "matter," or is it one of the first "parts" of the body to travel to unknown destinations? All indications are, the spirit IS the first to leave, as we hear from persons who have gone through a "near-death" experience during "clinical" death. Some individuals claim to have gone through this experience and had strange tales to tell after revival. Once again, it is very important to note, *NO* BIOLOGICALLY dead person has *ever come back* from the so-called "beyond!"

Looking at spirits (souls) from a scientific viewpoint, it leaves much to speculation and reliance on so-called "eyewitness" reports. As mentioned, many of the reports are either a hoax or a very strong belief in this phenomenon caused the observer to actually believe, he really saw what he THOUGHT he had witnessed. Other cases are simply very life-like dreams and the reporting person believed it to be a real life event. However, a number of cases are really almost impossible to explain away and much is left to speculation.

Typically, most encounters with spirits of the dead have been reported by one or two "witnesses" only and are therefore questionable. Encounters by SINGLE WITNESSES with ghosts or spirits, or for that matter with aliens or UFO's and without additional witnesses, are understandably received with much more scepticism.

An investigator, Hans Holzer, has been examining ghost sightings for a number of years. His success was mixed, but enough cases have been rather remarkable, when several subjects' reported sightings were investigated. Names, dates and places came up, which the subject could not possibly have known prior to the hypnosis session.

One such case comes out of from Port Clyde, Maine where a ghost turned out to be that of a lady whose husband vanished on a ship named "St. Catherine" in 1843. Under hypnosis, the subject reported correctly to psychic Ingrid Beckman that the "ghost lady" told her she had two children, one was a sailor during his adult life like his father, and the other one died at the young age of seventeen of cholera.

The ghost was said to be still waiting for her husband who disappeared with the whaling ship "(St.) Catherine." She claimed her name was something like Margaret Hatton, and that is was in fact 1843 and the sons' names were Philip and Frances. Most remarkably, all stories had later been verified through old records, together with the ghost's detailed description of the house, how it correctly appeared in the early 1800's.

Photography has greatly helped to overcome scepticism somewhat, when proven photographic evidence of ghost sightings can be introduced. Although many times such "evidence" has been faked, a considerable number of photographs have been lab-tested and were often found to be of "unexplainable" origin.

The best example coming to my mind, was a photo taken in 1922 near a London war memorial. After developing, the images of the faces of all fallen soldiers from the entire regiment appeared shrouded in a haze on the photograph. After scientific examination and to this day, no scientific explanation has been found to discredit or explain these and other photographic images of ghosts, and of course for comparison, those of UFOs.

Reports becomes much more creditable when a large GROUP of eyewitnesses have encounters with the unknown. Particularly, when the observers are in different locations at the same time and do not know each other. We DO have such reports on encounters with UFO's, or extraterrestrial spacecraft, but rarely if ever, with "ghosts."

As a final alternative to explain the existence of the spirit would be the earlier mentioned "aura" between living people and other primates. However, every type of radiation must have a power SOURCE and this theory would therefore only apply to the LIVING. A powerful family bond and willpower by one family member may be explained this way, when another family member experiences "mental telepathy" with the other. Some scientists have actually accepted the existence of an "aura" about LIVING persons, and it may even serve as explanation for the existence of a soul (or spirit).

This theory simply cannot be applied to the DEAD, since all power sources of a human cadaver are expired and decaying with the body mass. I leave it up to the reader to "believe" or "not to believe" in the existence of the ghosts and spirits of the dead. I am personally very sceptic about it.

Since the spirit is invisible during our lifetime, it is quite logical to assume that it must be also invisible after death, if it continues on and cannot appear as a "visible" ghost. Once again, one must leave it to scientific speculation or to religious belief, what the ultimate destination of our spirits is, if any. Do spirits stay around forever, or are they ultimately transferred into new bodies through reincarnation? Do they leave this planet to go to the biblical "heaven" or "hell?" Do they transfer to another planet into another living creature? The possibilities and speculations are endless.

Is it coincidence that as the human population INCREASES in numbers on Earth, some higher forms of animal species DECLINE in numbers or become extinct? Are the numbers possibly balanced out? One reality must be mentioned: nearly all religious AND many non-religious civilizations have been, and still are believing in one way or another in some sort of "life after death."

For those readers, who accept the existence of extraterrestrials, a strong belief in a transfer of "spirits" to another planet is also an acceptable theory.

Without scientific evidence of a "heaven" or a "hell," spirits may be leaving Earth entrapped in atomic matter, and could quite conceivably end up on another planet and solar system. At least, it would be somewhat scientifically logical. Maybe the spirits, which remain on Earth, are in "hell" and the others go to another better planet, to "heaven?"

However, one must keep in mind, the existence of an extraterrestrial planet as biblical "heaven" is physically impossible, conditions *can not* be "perfect" ANYWHERE in the universe. It is simply against all laws of nature. Hopefully, it may only be a few hundred years from now, when we will have the scientific, factual answers for this important part of our possible "after-life," if it exists at all.

REINCARNATION — REALITY AND A "TRANSFER" OF "SPIRITS"?

As I noted earlier briefly, a 100 percent reincarnation (or transfer of spirits) from one human life to another here on Earth is technically impossible, simply based on the following detailed calculations: The world population in 1988 was estimated to be well over five billion. Going back to the year 1650, we counted only 470 million souls, give or take a few hundred thousand. Where did the extra four and a half billion souls come from since that time? Whom were they reincarnated from? From other animals?

A significant "coincidence" was mentioned before, as the human world population increases over the centuries, many animal species become extinct or their number decreases just as fast! Therefore, if one accepts re-incarnation as a reality, for 100% of all humans AND other higher life-forms, only then the numbers may match, making re-incarnation a possible theory.

A number of creditable hypnotic sessions have been recorded where the subjects spoke foreign languages, described unknown ancient surroundings or wrote and read ancient Egyptian and Greek sign languages. Some subjects described the feelings and the cause of their death at the end of an earlier life. Dr. Wambeck of San Francisco specializes in mind exploration under hypnosis, probing for events in earlier lives. He had researched over 2,000 test "subjects," some of them reporting experiences going back to ancient Egyptian days. Frequently, the reports match with events described in history books, yet the subjects could not possibly have known about them for a number of reasons.

One such case researched by Dr. Wambeck was Ruth McGuire, who reported under hypnosis that she was known in France several hundreds years ago as "Maria, the healer." She was able to describe several verifiable events from her "past life" in great detail, including her death. Since (magical) "healing" was considered witchcraft at the time, she was condemned to die at the stake. However, she managed to flee and ended her life in a less painful way, by jumping of a high cliff.

At this time, we do have to leave reincarnation up to speculation, without scientific logic. One fact appears clear to me again, humans could not possibly ALL have been re-incarnated from previous lives. Similar with life after death as ghosts or spirits on this Earth, re-incarnation poses the same problem: billions of people have died over the past few hundred years alone. A few million people from pre-historic days could not possibly be re-incarnated into today's five and a half billion, no more than everybody living today could remain on Earth as a ghost or spirit. These are undeniable FACTS, and until we know the answers, one CAN revert to faith and religion only if it is comfortable for the believer. But it has to be done

WITHOUT having their own or other lives dominated by it, and only with greatest scepticism.

During a deep coma, we have no feelings and no emotions, although the body still functions and it is considered "alive." "Brain dead" accident victims on life support equipment are good examples. Several of these almost hopeless cases regained some or complete consciousness, but most patients could NOT re-call any feelings, sights or sounds from the period of their coma. Others claimed they were able to feel or even hear. What happened to the totally "lifeless" comatose victims' "soul" or "spirit" during the period of clinical death?

Although it may belong in the previous chapter, let me pick one example of a record length of medical clinical death: Ted Milligan of Winnipeg, Manitoba. He collapsed as a child during a frosty snowshoe walk. He was clinically dead and without heart beat for about two hours, until schoolmates finally found him. He has no recollection of the time of clinical death, no feelings, no pain etc., no evidence of a spirit after "death," or in-between life and death.

Evidence of transfer of certain emotions and "instinctive" movements and habits from our parents and ancestors is often noticeable, and appears to be passed on with other genetic information. Often a child walks in the same manner as one of the parents. Sometimes irrational behavior and mood changes can be observed to be parallel between parents and their children. A parent's habits, the emotional behavior and addictions are often found in children, even if the parent has long died or lived separated from the children for many years. Maybe THIS IS some form of reincarnation, at least it would account for some kind of transfer of spirit between living persons, along with the genes.

Every minute on this planet alone, an average of about 300 humans die or are born. Believers say that a "Creator," (God or Allah), is responsible for ALL deaths and ALL births and he would therefore have to decide over 300 souls a minute. Since it is an important point regarding reincarnation also, let us go over it again. I have committed more space to it elsewhere in the book.

I leave it to the reader once again to compare the physical impossibilities with religious myth. One can compare the idea of two to three "reincarnations" per second somewhere on Earth with the FACTS of the NATURAL events of birth and death. The answer should be obvious. Only a genetic transfer is remotely possible as covered in earlier chapters.

Eyewitness reports of "near-death" experiences from periods of temporary CLINICAL death, seem to include another phenomenon — victims usually reported a feeling of peace, first floating over their "dead" body and later drifting into a "tunnel" with a "bright light" at the end of it. Since we do not know, what to expect at the other end of the tunnel, we can BELIEVE in a number of possibilities, without accepting them as fact. The easiest way out, but not at all logical, is to believe in the scripture: "We will all come to our maker and be judged by our behavior on Earth during our life time." This explanation is not satisfactory to all the sceptical millions, again simply because of numbers. How could "God" judge EVERY dying person and "create" EVERY life, there is simply not enough time, no matter how much we believe in the capabilities of a superhuman "God." But I have gone over this before in detail.

I brought up the possibility of reincarnation on another planet within a distant star system before, at least it COULD be a more logical explanation with a somewhat scientific background. The reader will have a look at the immense number of star systems in our own galaxy, the Milky Way, in the chapters on astronomy. It was estimated during a conference of astronomers as early as 1961, that the existence of extraterrestrial, more or less intelligent civilizations in our galaxy alone may possibly be as high as 40 to 50 million, or as "few" as 50,000.

Only one century ago, when the universe was thought to be consisting of the visible stars and planets only, this theory would have been totally unacceptable. Early scientists had to be very careful to proclaim this kind of fact-based speculation, since they had to be afraid to be ridiculed by religious fanatics, whenever scientific theories clashed with the biblical teachings.

Maybe the estimated high number of forty to fifty million potential extraterrestrial civilizations is generous and most of them are possibly LESS advanced than today's earthlings. But if we cut the probability to 1/1000th of 40 - 50 million, which was considered the most "conservative" estimate, we are still left with 40,000 to 50,000 highly advanced civilizations. Therefore, it is more logical to "believe" in reincarnation on another planet, if one considers reincarnation a definite option at all.

A perhaps somewhat believable theory about the "bright light at the end of the tunnel" is that the "spirit" of a clinically dead patient is about to enter a new life during childbirth under dream-like brain activity. I.e., the bright light at the end of "the tunnel" is thought to be from the lighting in a delivery room during child birth. The theories are endless, but all are only *theories* without scientific proof and usually without logic. Of the billions who have died over the past 2,000 years alone, nobody has ever come back to tell us a "factual," or at least a believable story about re-incarnation.

EXTRATERRESTRIAL VISITORS, NO LONGER JUST FICTION, THE EVIDENCE IS MOUNTING

July 11, 1991 over Mexico City, during or about the same time as a total solar eclipse: Thousands are watching this once-in-a-lifetime experience, a total eclipse right overhead. the camera crew of the Mexican equivalent of "60 Minutes" had camera equipment rolling to record the eclipse. At the same time, a UFO was spotted and at first filmed more or less by accident. The usual familiar disc-shaped forms were clearly visible as the object was moving slowly in various directions for a considerable amount of time. Thousands witnessed the movements and the eventual quick departure at speeds, unattainable by human technology. Although the incident was reported in detail by the Mexican media, little was mentioned about it in the rest of the world.

Millions believe in past or present alien visits to Earth or they accept it as undisputable "fact." It is no longer a question for me and over fifty percent of the North American population, (according to several polls) of merely BELIEVING in UFOs. Other chapters also deal with it, since too many creditable eyewitness reports confirm their highly probable existence. Firm SCIENTIFIC evidence is either non-existent or it is suppressed by governments. I will go into this controversial subject later. Also in this context, I want to revert back to the chapter on Adam and Eve, why has "God" chosen the HUMAN body for us and for himself, on this Earth one of nature's most imperfect bodies?

Why did "He" not select a more intelligent, more rational, more peaceful and physically more advanced creature, as they may be found on OTHER planets, or for that matter, on Earth right now? As we go back into "biblical" history, man's body was by no means a better "selection," and if anything, people also had a more irrational, primitive mind than today.

Evolutionary developments apply to all mammals, and some of are them long extinct under the law of survival of the fittest. It appears, that entire human races also have disappeared for the same reason. But some of them DID survive, and we are here today as living proof. The laws of nature and evolution must work in the same way on other hypothetical planets, where life and intelligent, higher life-forms have developed. Conditions for the formation of higher life forms would have to be comparable to somewhere between subtropical conditions on Earth 300 million years ago, arctic climate today and in the oceans past and present. Those were and are the conditions for life as we know it to form and to be sustained.

As the atmosphere and climatic conditions on Earth were entirely different hundreds of millions of years ago, so could the atmosphere be different on other planets and still encourage the formation of higher life-forms, AND sustain them for millions of years. The equivalent to homo sapiens could develop elsewhere, if the alien's physical and mental activities were much like those of mankind. The fact is, all chemical elements appear to be the same on all other planets, only the PROPORTIONS vary greatly along with the climate. Therefore, I can find no reason why primates very similar to those on Earth, could not develop elsewhere.

Alien primates would be subject to changes in physical appearance through evolution in accordance with the laws of nature as we are on Earth, past, present and future. Hence, they may quite possibly have the exact appearance as it is often described by terrestrial eyewitnesses from so-called encounters of a third or fourth kind.

Unless there is a major natural or man-made disaster on Earth or any other earth-like planet, the evolutionary process will continue until our Sun (or theirs) comes to the end of the natural life cycle, during the final stages of "main sequence" (see the chapters on astronomy). By that time, all intelligent life will have left the "home planet," since NO life form will be able to adjust to the dying stages in a planetary system's life. More about it in the chapters on astronomy.

Let us look again at the history of humanity and other primates on Earth. Mankind has been and still is more often than not, too desirous to kill one another during armed conflicts, religious, territorial or otherwise motivated. The earlier inability to prevent epidemic diseases contributed equally to a rather slow initial growth of the Earth's population. The numbers could probably be four, five times today's totals and scientific knowledge could have been much further advanced, had scientists being able to concentrate on peaceful projects.

Increased population would NOT necessarily have caused mass-starvation for millions, since scientific knowledge could have made early use of still unpopulated parts of the world . But unfortunately, our ancient ancestors AND our more recent forefathers have not used their brain-power to contribute to priorities such as effective farming through artificial irrigation, just as one example. Instead, their priorities were directed towards more efficient ways to conquer, kill and destroy, and to forcefully introduce mythical religions.

Believers should question why their almighty "God" could have allowed hundreds of millions to die during senseless religious and territorial wars over the past few thousand years. Humanity created their own conditions and circumstances on Earth, one can only hope that humanoids on other planets were wiser.

Back to the main subject of this chapter — extraterrestrial visitors — with more detail and more emphasis on some of the recorded FACTS: Unidentified flying object have been observed and photographed so many times by so many reliable eyewitnesses that one must consider some of them to be REALITY. One of the latest estimates is that about 85% of "sightings" and "encounters" are explainable as terrestrial events or hoaxes, leaving us still with 15% of all reports to be "unexplainable."

Famous scientists, including Dr. Carl Sagan and Dr. Isaac Asimov may want more hard evidence, before they are convinced about the existence of extraterrestrial intelligent life. In others' opinion it is no longer only a matter of just "believing" or "not believing" in UFO's. Many of the encounters can simply not be ignored. I will take a number of selected examples at this time, since it would fill several volumes alone to list ALL reasonably creditable sightings. Much has been written about this phenomenon, but too often obvious hoaxes and real encounters knowingly had been mixed to represent "facts," or to create sensationalism for supermarket tabloids.

Most convincing alone is that until a few years ago, over 3000 sightings had been reported by commercial or air force pilots. Amongst many others, the Russian fighter pilot Maxime Cherbacov reported a sighting. During one of the 1991 NASA shuttle missions, the on-board astronauts saw and recorded a UFO sighting. Surprisingly, this event was reported during a rather obscure TV program, while one would have expected this kind of news during the evening reports on national television.

As recent as March 8th, 1994, a police officer and dozens of creditable observers reported a very large object over Holland, Michigan. Radar operators in the area also observed the object on their screens and recorded their own conversation on audio tape, while the UFO was clocked at moving about 3600 m.p.h., at less than 6000 ft. altitude. Some observers thought the object to be very large, others guessed it to be MORE than one craft, possibly two or three flying in close formation. However, all agreed on the actual time and place. As yet, no man-made object can travel at THAT speed while flying at THIS relatively low altitude.

Shortly after leaving Melbourne, Australia on December 21, 1978, the pilot of a single engine plane reported the sighting of a UFO. He claimed to have seen green lights coming from it, and he was sure it was not another aircraft. Also, no other aircraft traffic was scheduled to be in this area at the time. The UFO was to have circled his plane and at last report, was right above him. The object was picked up on radar off and on, but the irregular speed had the air traffic controllers confused.

The pilot reported engine trouble and shortly after, his plane disappeared from radar without a trace. The UFO also vanished. The event was filmed by other eye-witnesses, as they spotted the plane and a mysterious green light, both coming together very closely. They also witnessed the disappearance of the plane and the UFO beyond the horizon.

The plane, the pilot or remains of either have never been found, and the Australian Government has never released the audio tapes containing the air traffic controllers' conversation. During the late seventies, UFO sightings were reported frequently over various parts of New Zealand, including some daytime sightings.

Encounters with UFOs, (alien spacecraft?) have been reported by about 15% of all airline personnel world-wide. The American scientist Ted Phillips has examined over 1,400 landing sites, where physical evidence had been left behind. Most of them appeared genuine, only a few may or may not have been hoaxes. In contrast, how much scientific evidence do we have to back up religious myth?

UFO's sightings have been recorded as far back as 58 AD by a Roman soldier. He reported an erratically, very fast moving disc shaped object in the sky. Of course, at that time, Earth's inhabitants had no idea what it could have been. There are many more recorded ancient sightings and one in particular stands out in my mind: on November 1st, 1461 a UFO sighting by thousands was recorded over many parts of Europe.

A sighting in Nuernberg, Germany is recorded on a painting dating from the year 1561. Records of a sighting in 1566 over Basel, Switzerland have been found.

One can actually find one example of a sighting in biblical records. Ezekiel wrote about the incident as "a wheel in the middle of a wheel" (Ezek. 1:16). This strange spacecraft was to have been able to move in any direction without turning, it could change instantly from one direction to another. Also according to the Bible, "Elijah" was picked up by a "chariot of fire and horses of fire, and Elijah went up by a whirlwind into heaven." He was said to have reappeared in the same manner.

So far, and if this ancient biblical story is reasonably correct, the report sounds exactly like a modern-day encounter of the fourth kind. Some television evangelists now have adopted these and other biblical sightings with less scepticism, since "it is written in the scripture." Any RECENT CREDITABLE reports may be classified as hoax or examined with much more scepticism by the same people, if it does not fit into their interpretation of the Bible.

Some television evangelists use selected reports of UFO sightings now as an interpretation of visits from the biblical "heaven," ultimately to be by Jesus Christ. How does THAT agree with the biblical myth that humankind was created on Earth, the hypothetical centre of the universe, and in the form of Adam and Eve? If religious leaders except ONLY ONE ancient alien visit today as fact, other than "God's," are they not admitting to the existence of OTHER intelligent extraterrestrial intelligent beings?

In this context, one may note from modern day believable eyewitness reports of encounters of the third or fourth kind, no such visitors have ever been reported to have acted in any manner, that could imply a visit from "heaven." The visiting aliens apparently never left the impression that they are from "heaven" or that they are sent by a God. Or that they ARE gods.

What we CAN conclude, is that they are technologically extremely highly advanced, and that they come from planets with a much more advanced evolutionary process. The fundamental requirements for life appear to have existed much longer and possibly in some cases are more favorable on their planets than on Earth. Even a few hundred years would make a tremendous difference to technology, and a few hundred thousand years of evolution and scientific discoveries will change any planet.

It is also notable, alien visitors and their spacecraft are not always described to be identical in appearance. Therefore, it is logical to conclude that extraterrestrials are coming from *different* interstellar planets and not just from one distant solar system. Although visiting aliens are expected to be technologically greatly advanced, they are all subject to the same laws of the universe and those of physics in general.

They may possibly live much longer, may have overcome many earthly diseases, and seem to have found ways to manipulate gravity, but they would still have to eat, drink and digest food products. They would have to protect themselves from the temperature extremes of interstellar space and dangerous types of radiation. Light for advanced aliens also travels at 186,000 miles per second and all other laws of physics and chemistry apply equally. Only, they seem to have overcome some of nature's obstacles long before us.

A TV evangelist's imaginary "heavenly" visitors obviously do not fall into this category, they seem to be beyond all laws of nature. Yet, God and Jesus Christ are portrayed as "human" in appearance. "It is written" that Jesus ate and drank like the rest of the locals during his days, yet he could "rise from the dead" and disappear from Earth without a visible spaceship.

Religious fundamentalists and firm believers just simply overlook the obvious by taking EVERYTHING that was written by a small group of locals three and four thousand years ago for granted. And by accepting these same ancient writings without scepticism. Obviously, so much was lost in the translation that leaders of major religions cannot even agree on what REALLY happened, hence religious conflicts and wars.

Getting back to this century and to our capability to explain the appearance of extraterrestrial space crafts scientifically, here are some more facts: Prof. Stanton Friedman of California has investigated UFO's for over 25 years. He reports speeds of about 10,000 m.p.h. and spacecrafts making sudden ninety degree turns at this speed, left or right, AND up or down. An impossible task for even the most advanced human technology of the nineties.

Colonel Wendell Stevens of the U.S. Air Force observed a UFO over Alaska in 1947 and measured speeds up to about 7,000 m.p.h. Later as a U.S. Air Force General, he received reports of sightings on an average of twice per months.

The first "flying saucer" was observed over the state of Washington, also in 1947, from a small plane during an unrelated rescue mission. The description "Flying Saucer" origninates from that date. 1947 was a very active year for UFOs, as we will see later in this chapter.

Former President Carter, while he was Governor of Georgia, and a number of his guests, observed a UFO. Or was it Venus that he saw?

A Navy photographer by name of Hansen photographed an alien spacecraft over Catalina Island in 1966. He estimated the disc-shaped object to be of about 30 ft. in diameter. Other reports also give a 30 to 50 ft. diameter, round and saucer-shaped, nearly silent and well illuminated during night sightings.

The controversial Travis Walton incident of Arizona was reported in November 1975. Six men witnessed the landing of a UFO. One of them, Travis Walton, became a bit too inquisitive and ventured closer to the object. He then claimed that he was abducted, as witnessed by his friends, and he later had disappeared for five days. Before his return, his partners were accused of his murder and had to undergo separate lie detector tests. They not only passed the tests, their stories were as good as identical. Under hypnosis, Walton later reported of strange tests which the aliens performed on him during the five day ordeal.

Lie detectors have been used on eyewitnesses at other occasions of UFO reports. A Mr. Crichfield had sighted a UFO near Big Chimney, West Virginia, on June 12th, 1975. Needless to say, he was questioned by sceptics and his story had to be confirmed through a lie detector test. The test proved beyond a doubt that he was telling the truth, as both stories were identical — before and during the test. The University of Kansas tested samples of the burned soil at the landing site, and it was concluded that it was not a man-made incident.

While on the subject of burnt soil: Janet Kaye of Medford, ME, saw and reported a UFO sighting, in November 1975. Many other townspeople also saw the same alien craft, which after departure had left a large area of burnt grass and soil.

In Bentwaters, England, physical evidence of a landing was found by a number creditable witnesses, but it was immediately confiscated by officials.

Probably the most famous incident happened in July of 1947 near Roswell, New Mexico, when a number of witnesses were surprised by a significant crash of what they thought, was a plane. Upon closer examination of the wreckage by businessman Dan Wilmont, Jess Marcel and others in the immediate vicinity, it became quite clear that the object was of extraterrestrial, or at least totally unknown origin.

The material from the wreckage was of a completely unknown substance. Some pieces were marked with an unidentifiable writing. Other material appeared to be as thin as tin foil, yet the witnesses were unable to bend it even with a sledge hammer. Captain Henderson saw alien bodies amongst the debris, but he had been quiet about it until recently.

After the wreckage and the bodies had been moved hastily from the site to an Air Force base about 600 miles away to compound area 51, the mortician Glen Dennis had a call from the base inquiring about the availability of small, child-size caskets. When he visited the base upon invitation there appeared a lot of confusion amongst the personnel. He noticed doctors and nurses had been overcome by noxious fumes and he later had a secret meeting with one of the nurses.

At the time, however, no one wanted to talk to him and he was abruptly escorted from the premises. He was told in no uncertain terms that he saw nothing and he was verbally threatened if he should talk about it to anyone. Incidentally, the nurse who had described the three bodies to the mortician during their meeting was later involved in a fatal air crash.

The initial and official press reports were referring to the wreckage of an "unidentified flying object." Only one day later, the report was retracted and the wreckage was explained as that of a "weather balloon."

To this day, no official explanation, other than the "weather balloon" story has been given. We may never know from official sources. The question is — if it was not an aircraft and not a weather balloon and or some secret weapon (with three-foot pilots in it?), what was it? All evidence points to a malfunctioning extraterrestrial spacecraft.

Pentagon employee Larry Bryant tried to investigate this incident on his own and formed an organization named "Citizens Against UFO Secrecy." He took the U.S. Government to court under the "freedom of Information Act," but since he had no proof of the actual location of the wreckage, the case was dismissed. To this day, he has not given up and is working on another court action, all according to a recent TV interview.

It does prove one point, even the most unimaginable technically advanced civilizations from far away places still have accidents and mechanical failures, they are victims of circumstances (gravity), just as earthlings. Years ago these visitors would have been looked upon as "Gods." But then, three-foot gods do not have mechanical failures and do not die as a result of them, do they? In other words, there are no "superior beings" known to us as yet, who are NOT subject to the laws of physics and of the universe in general.

In the Roswell incident all physical evidence was reportedly removed by the U.S. Air Force and ultimately transported to the Wright Patterson A.F.B., Hangar 18. The incident was always categorically denied. Yet, all witnesses swear to it that the material from the wreckage, which they had touched and held in their hands, was nothing at all ever to be found or manufactured on this planet. Like the "tin foil" that could not be bent or dented with a sledge hammer. Others, including the former Senator Barry Goldwater were denied entry to Hangar 18. How much more evidence do we need about the existence of intelligent extraterrestrial life?

Here is more: Three individuals in Wisconsin, including Sheriff's deputies witnessed and videotaped a UFO on February 4th, 1993. They were able to track it for about twenty minutes until the object disappeared.

The next item alone should sway a few sceptical readers from being just "believers" into accepting alien visits as "reality."

I am referring to a full-page article in a reputable Toronto daily newspaper, not a super-market tabloid, out of the eighties. Roy Bonisteel, the host of the CBC (Canadian Broadcasting Corporation) series "Man Alive," had covered a number of spiritual subjects in over twenty years. He can be considered a most rational and open-minded individual, if anybody can be. Roy Bonisteel and the series' producer, David Cherniack, B.Sc., did considerable research on the subject of extraterrestrial visits, in particular on "encounters of a fourth kind."

In cooperation with Budd Hopkins, Dr. Kaufman, astronomer Terrence Dickinson, Allen Tough and John Musgrave, a number of reported encounters were discussed and the results of 120 interviews and findings from hypnotic sessions were analyzed. Descriptions of alien abductions given by "victims" from across North America all appeared consistent. The abductees' reports on language and technology seemed nearly identical and many encounters coincided with UFO sightings by others in the reported areas. All victims claimed to have lost a period of time, from when they have no memory. Only hypnosis brought the rather frightening experiences back to their minds.

Bud Hopkins had studied the subject for over 18 years and wrote two books about his findings. Many of the abductees interviewed by him were unwilling to talk about their experiences, fearing ridicule. In general, and in addition to a memory loss for hours or days, all recalled being unable to move while they were examined by the aliens on some sort of operating table.

During the meeting of these experts, one or two encounters were particularly highlighted. A lady by the name of Dorothy and another by the name of Betty claimed to have been "worked" on in a barn owned by Betty. Neither women ever knew about the other until Dorothy was taken to the barn by the aliens. Dorothy's description of the barn under hypnosis matched the real thing. Dorothy, like a number of other abductees, had more than one encounter of the fourth kind, she claimed to have been first artificially inseminated, and during a later visitation, the embryo was removed. At a later date again, she was shown a rather sickly looking half-human, half-alien creature, which was explained to her as her off-spring. There was some physical evidence to back up her story at least partly, such as a patch of burnt Earth on her rural property, where grass has not grown since, and a ring-shaped marking on her leg, similar to other reported encounters.

The case of Betty and Barney Hill was discussed again, the New England couple who experienced memory loss, similar to other encounters with aliens of matching description. Betty Hill was able to sketch a star chart from memory, describing the home star as "Zeta Reticulum", which is a binary system (twin star), and according to our scientists not really a likely place for stable life-forms to develop through an earth-like evolution. Planets in binary systems are very unstable in their orbits, due to the powerful gravitational forces from TWO solar sources. However, the chart drawn from memory by Betty Hill gave a reasonably accurate location of the star system, considering that the entire experience was revealed under hypnosis.

The CBC group summarized their thoughts to the point, that it will take a radical re-evaluation of humanity's age-old picture of our culture and religion. Bonisteel's comments were: "What happens to our culture and religion, to our very sense of ourselves as a race?"

People who find solace in religion, says Dickinson, would be forced to ask the "Almighty": "Who were those little guys in the parking lot and are you their God, too, or what?"

The group rightfully concluded that scientists, who do not already believe in the existence of extraterrestrial life, should examine the reports more closely and try to explain them. Or as I want to add, TRY to explain them ALL AWAY or write them ALL off as hoaxes or nightmares. I think it is time now for people everywhere to look MORE sceptical at old religious writings and a lot more open-minded at the reality of existence of other far more intelligent life outside the solar system.

Getting back to other reports and signs of possible much earlier alien visits; the French scientist Maurice Chatelaine has discovered that twelve ancient temples around the Aegean Sea are arranged to form a perfect giant Maltese Cross of about 300 miles in diameter. It is only visible from great heights and would be meaningless to the ancient civilizations on the ground — is this arrangement really just coincidence? It has been calculated, if the placement of the temples had been at random, it would have taken 95,000,000 years of trying to come up with this exact lay-out! Is it not much more logical to accept it as fact that it was designed and the temples were built, or instructed to be built, by a highly advanced extraterrestrial civilization? One point is clear — the answers cannot be found in the scripture.

The "SETI" program (Search for Extraterrestrial Intelligence) is operated by various scientific institutions, and until recently recorded ten million times ten million mostly explainable radio signals from outer space. The exceptions are thirty-seven signals, for which there is no explanation.

A cooperative program to SETI has been in operation since April 1992 — "SERENDIP III," for "Search for Extraterrestrial Radio Emissions from Nearby Developed Intelligent Populations." By late 1994 an incredible 70 trillion channels of data had been analyzed on location at the University of California at Berkeley, yet only 30% of the sky had been searched. Every 1.7 seconds a total of 4.2 million channels are analyzed, and the program has come up with 400 signals which may be considered "candidates" for further examination. SERENDIP IV will be operational by mid 1995 and will be able to increase the search factor further by 40.

Another program, META (Megachannel Extraterrestrial Assay) is operated by Professor Paul Horowitz of Harvard University. He has so far detected 56 signals that also could be interpreted as alerts form extraterrestrial intelligence. The reader may want to remember, that any real or imagined messages from other star systems would have to take several years to get here, since no star is closer to Earth than about four light-years. Therefore, it would take radio waves at least that long to get to our receivers, travelling at the speed of light.

Getting back to Chatelaine, he also writes about discoveries in Iraq as proof of alien visits 300,000 to 600,000 years ago. He wrote a book on the subject of the origin of modern man. Based on his research of scientific findings, he believes that modern Homo Sapiens originates from a mixture between primitive, prehistoric men from the evolutionary chain on Earth, AND was interbred with highly advanced extraterrestrial visitors, starting about 60,000 years ago. Modern man in the present form appeared around 35,000 years later and had inherited only some of the physical features from the prehistoric humanoid creature, which had been around until about 60,000 years ago.

The physical features of early modern man changed somewhat radically in the relatively short period between 60,000 and 25,000 years ago. His theory is about the most logical and most acceptable, as I see it.

Chatelaine further refers to three pre-historic temples in France, separated by hundreds of miles, but lined-up in a perfect triangle of equal distance between them and at exact proportional dimensions with the Egyptian Cheops Pyramid. Again, no earth-bound primitive men could have come up with this perfect geometrical arrangement. To make it even more interesting, most of France's uranium mines seem to be cluttered along the lines of the triangle. Is it coincidence or did alien visitors need uranium to fuel their spacecrafts?

It has been estimated during a recent broadcast of a TV documentary series that approximately twenty-five million people around the world had reported either a sighting or an encounter of one kind or another with UFOs or aliens. In my estimate, if we consider 50% to be illusions or hoaxes, it still leaves 12.5 million encounters. Let us be extremely pessimistic and assume that only one percent of all reports are to be taken seriously. It still leaves us with a respectable number of 250,000 creditable encounters — can we dismiss such numbers ALL as false or imagined? Not likely, if we are realistic.

On the same program, the Pulitzer Prize winner and psychiatrist John E. Mack discussed a number of encounters by abductees and he made it clear that he is convinced the reports are not some "vague symptoms of presence," in his words. He fully believes in most of the reports he had investigated. The question came up: Could the alien visits be from another, as yet unknown dimension and are the abductees temporarily part of this other dimension?

The possibilities are endless, but nothing is to be found in the scripture that could explain any of this to us. Unless we put God and the "angels" into the category of alien visitors from another dimension, being just SOME of many such visitors?

The psychologist Robert Becker does not agree with Dr. John Mack's findings and he contributes the abductees' reports as originating from a half-asleep, half-awake type of "sleep paralysis." It may be so in cases of one individual reporting an incident, but not likely when *several* persons report details on one and the same abduction. A number of abductees would have to experience the identical event and talk about it under hypnosis.

Natives of all continents have been looking upon visitors from outer space as "Gods," as far as we can trace history back. Our ancestors before modern religions obviously admired and worshipped the advanced intelligence of the visitors and in some cases they are still waiting and hoping for the return of the gods. However, the ancient "Gods" were usually not considered "Creators," contrary to current religions. The universe and the smallest particle as such, was and still is, the source of life in many native legends. I have been referring to examples throughout the book, such as the discovery of ancient drawings of spacemen-like visitors in prehistoric caves.

Most amazing to some archaeologists was the discovery of SHOE prints, (NOT bare foot prints) in Utah together with 400 million year old remains of crabs, imbedded in the same stone. A hoax, a coincidence? Not likely, only religious fundamentalists may wish it was. Is it not simply possible that technologically highly developed aliens from other worlds explored a hostile Earth about 400 million years ago and left these prints?

More sightings to come: An object crashed near Keksburg, PA., in 1965. Over sixty eyewitnesses reported the event, but the wreckage was quickly removed by the U.S. Army for unknown destinations.

On April 17, 1966 Deputy Sheriff Dale Spaur in Portage County, Ohio, together with another deputy, investigated an abandoned car. Out of nowhere appeared "this thing as big as a house, bright enough to make your eyes water." They were then ordered by radio from headquarters to give chase and ended up following the object for more than 70 miles, at speeds of over 100 miles per hour. After a second cruiser with two more police officers was included in the chase they all ended up near Conway, PA. All four officers saw the UFO shoot straight up and disappearing at an incredible speed.

Also in Ohio, to be exact over Mansfield on October 18, 1973, Detective John Healey and three other helicopter crew members spotted a UFO near their aircraft. A powerful green light beam from the UFO altered the chopper's course so much that with the on-board equipment in descend mode, the aircraft actually ascended uncontrollably. All instrumentation went out of control and even after the UFO disappeared after a number of scary minutes, the equipment was still most unreliable. The helicopter was retired after this episode, since the instrumentation continued to malfunction and was not repairable. The Delong family also observed this encounter independently from their car.

On Aug. 7th, 1970 over Saladare, Ethiopia, at about 11:30 a.m. a red glowing object appeared and swept over the village, leaving a path of destruction. During a ten minute flight over the village, houses were damaged along with stone walls, trees were up-rooted and asphalt, as well as cooking utensils, melted from intense heat. Yet, no fire ever started throughout the ordeal.

In Scandinavia and Rumania during 1972, several mysterious radioactive impressions were discovered, after strange objects had been seen landing by eyewitnesses. In the Rumanian incident, a perfect circle was left in a cornfield together with a hole in the ground, six inches around and an amazing eight feet deep.

Quite recently, the former KGB files on the UFO phenomena were released to the Moscow correspondent of the ABC network. Multiple sightings are recorded on film or video and sightings of UFOs' as large as 900 ft. diameter are included. Hopefully we will hear and see more about it soon.

On November 2nd, 1957 near the little town of Levelland, Texas, Patrolman A.J. Fowler received calls from seven different motorists. With their identical reports, all had observed a large glowing object either hovering above or stationary on the road. Their car headlights had gone out and car's engines had stalled. Shortly after that, five more police officers saw the same object.

Also in 1957 over the coastal waters of Brazil, an explosion of an unidentified flying object was observed by several eyewitnesses. A search later revealed three tiny chunks of a light metal, which were later analyzed in Brazil and in the U.S.A. It turned out to be a nearly pure form of magnesium, which lacked the usual earthly elements. The pieces contained strontium, not normally found on Earth and equally uncommon in earthbound meteors. However, the U.S. Air Force investigators insisted that the fragments were indeed of meteoric origin.

Again in 1957, a young Brazilian farmer, Antonio Villas Boas, claims to have been taken aboard an alien, egg-shaped spacecraft by three of its strange "passengers." In what sounds like pure science fiction, he claims to have been stripped and taken by an alien nude woman of about 4 feet 5 inches in height into a separate cabin, where he was to have been "raped." However, a Brazilian doctor examined the victim and claims to have found strange evidence of a physical encounter, such as scratches and marks, and more important, indications of exposure to a high dose of radioactivity. If everything else was fabricated, where would he have picked up the radiation?

The next encounter involved more than one witness, to be exact, thirty-eight of them. In the evening of June 26th, 1959 at the Anglican mission station in Papua, New Guinea, Father William Gill and other, mostly native witnesses noticed a "sparkling object" above. It was reported to be hovering at a height of about 300 to 400 feet above ground level, a circular-shaped craft with a wide base on four legs. On top of the alien spacecraft, four humanoid figures appeared and seemed to make some observations of their own, they acted as if they were "adjusting" some equipment.

At one point, one alien looked at the frightened earthlings below and answered a waving motion from Father Gill by waving back. Shortly after, all four aliens were waving to the earthbound observers. The following evening, the craft appeared once again and stayed for several hours. After darkness had set in, Father Gill signalled to the aliens with a flashlight, and to his surprise, the UFO pilots replied by sending similar, but unrecognizable signals.

Sheriff's deputy Arthur Strauch, his wife and two friends photographed and drew a picture of a UFO in St. George, Minnesota in 1965.

Four Army officers nearly collided with a UFO during a routine flight near Columbus, OH in 1973. At least one known witness on the ground observed the same craft.

In California during 1975, Joe Lugo, his wife, their daughter and a niece saw and drew a sketch of a UFO near their home. An investigator from the North Western University's Centre for UFOs was convinced about the sincerity of the report.

Finally, during a camping trip in Maine in 1976 near a sacred Indian site, Jack Weiner and two friends observed a UFO. Later his friend Chuck saw the object again from his canoe. He flashed an SOS signal to the alien craft with his flashlight and to his surprise received some sort of reply from the alien craft, in form of light beams. Soon after, the UFO came closer and what appeared only to be a short time later, it had disappeared again. The observers later discovered a time loss and all were very tired.

So far, this episode could have been fabricated. However, after Jack Weiner repeatedly had nightmares much later, all four went through hypnosis 12 years after the encounter and during individual sessions revealed an abduction by aliens in 1976. The results were examined and found to be identical. All four also passed polygraph tests with the same results.

I purposely only picked a few of the most convincing reports. Many books have been written about the existence or non-existence of UFOs and their extraterrestrial origin. Anyone may take all reports at face value or as fact. Anyone, who wants to categorically deny their existence will also find enough reasons to dismiss ALL sightings as hoaxes or as illusions, optical or otherwise.

Personally, I accept their existence with a certain amount of scepticism, and I agree, many sightings are unrealistic, maybe even MOST of them. But with the sheer number of reported sightings by sincere professionals, by public servants and officials, Armed Forces personnel and by independent witnesses from different observation points at the same time, a respectable number of them really do deserve to be accepted as reality.

Remarkable discoveries of ancient visits from space have been found in many places by archaeologists. A "man-made" sparkplug-like device was discovered by rock hunters in the south western U.S., imbedded in a 500,000-year-old rock.

With the number of creditable reports on file, one may begin to wonder why these stories did not make headlines, or at least front page in the daily newspapers? Usually the reports (creditable or questionable), can only be found in tabloid papers.

American Hopi Indian natives believe they originated from another world. They display dolls with the appearance of strange looking "astronauts" and they have a collection of small replicas of saucer-shaped UFOs.

Other natives in the American West display paintings in their caves, of animals and spacemen together. Some believe in a legend about two spacecraft, which had landed many years ago. One had crash-landed and the second one brought some modern electronic equipment, possible for repairs. The Indian people painted the display images of oscilloscope screens, a modern electronic device, on the walls of their caves. Sine waves, sawtooth waves and square waves were totally unknown to modern man until this century

Gigantic stone blocks of unknown origin weighing about 800 tons each can be found at the location of the ancient temple of Baalbek, Lebanon, at an altitude of 3,800 feet. It would be impossible to transport single objects of *that* weight to *that* altitude with modern technology.

The remains of a huge figure of a man can be seen in the Mohavi desert from a plane only.

The Egyptian pyramids are arranged in a perfect line-up as markers for landing by ancient visitors from the sky, at least so it appears.

The African Dototh tribe believes man lived originally in the "sky" with the god "Akuj," (in an alien space station?).

Many readers may be familiar the markings in the plains of the western Peruvian Andes mountains, also known as the "Lines of Nasca." Along with replicas of modern aircraft runways, huge figures of spiders and nine-fingered monkeys are visible from the air only, covering an area of about 200 square miles. The drawings make no sense whatsoever to the observer on the ground.

A 24 ft. high monolith in Tiahuanacu, Bolivia, dates back to the tenth century in honor of the "Great white God," who was to have visited the natives at that time with the promise to return. When the Spaniards arrived hundreds of years later, they were mistaken for the returning "White Gods." As a matter of fact, the early alien (white) visitor and the Spanish were both named as "Viracocha." It turned out to be a fatal error for many natives.

That brings me back to religious teachings. Since the

natives of Bolivia had no knowledge about the existence of the biblical Middle East, "our" God and their "God" could not have been one and the same. At least, nothing in the Bible covers the (biblical) "God's" visit to other people on the other side of the then unknown Atlantic ocean. The biblical Middle East WAS the whole world.

Guatemala's 4000 year old statue called "Fat Boy," may only be six foot tall and would be possibly hard to detect by extraterrestrial visitors from the sky. Yet it is worth mentioning since it has magnetized stone at the navel and at the temple of the statue. Where did it come from?

More strange markings: A pile of rocks shaped into the form of an eagle with a wingspread of about 120 ft. can be seen on top of a mound, estimated to be 1500 years old.

At Uffington in Britain, the figure of a 360 ft. long horse can be seen from the sky for the last 2000 years. Again, it only makes sense, when it is observed from great heights.

The Great Serpent Mound of Adams County, Ohio is one of the largest markings in North America. At 20 ft. width and 5 ft. height, this "snake"-shaped stone monument is about a quarter of a mile long.

Getting back to Peru, on the hills facing the Bay of Pisco, markings are visible again from the air only. A National Geographic study confirmed an astronomical significance to these lines.

In the Museum of Aeronautics in Lima, unusual ancient stone structures are displaying drawings of strange flying objects and what appears to be pictures of some sort of astronauts. The Museum directors decline to comment on this possible extraterrestrial interpretation of the drawings; after all, it would not comply with their strong catholic belief.

The same goes for publicity around Father Crespi's Museum in Cuenda, Ecuador, where a heavy solid gold plate is displayed with as yet undeciphered hieroglyphics; ancient Egyptian writing in South America? It seems, nobody is too anxious to decipher this text. Maybe the good father could find out something that is not in compliance with catholic teachings?

The "Stonehenge" in England was built approximately 2000 to 2500 BC at the same time, when some of the Egyptian pyramids were built thousands of miles away. The hard, blue rock used for the Stonehenge was transported from the "nearest" source in Wales about 240 miles away. Astronomer Dr. Hawkins has calculated that the Stonehenge represents the hub of a vast grid-like network of man-made rock formations all over the southern UK. For what and for whom?

The largest, nearly 2000 year old pyramid on Earth is not located in Egypt, it can be found in Mexico covering an area of about 45 acres. Were the architects of it the same as those of Egyptian pyramids? If so, what were their means of travel at a time, when the Middle East by some was considered to be the entire world? If they were not the same architects, how did the builders communicate with each other? An ancient replica of the Stonehenge has been found in central America, the so-called "Woodhenge." Once again, who built it and why? Surely not for the Judea-Christian-Islamic God!

Large flat and mainly square rock formations were discovered about twenty years ago in the waters near the shores of Bimini Island, Bahamas. The rocks appeared to be a sunken wall or some sort of road. Further discoveries revealed collapsed building columns and a sunken pyramid. The similarity between the sunken building debris off Bimini, and that of ancient Greek and Egyptian pyramids and other structures, is remarkable. It leads to the conclusion that all of it was built by the SAME civilization and at about the same time in history.

While building similar structures thousands of miles APART about 2000 to 3000 years ago, how were these civilizations in contact with one another? Without modern long distance communication systems and without the knowledge of each others existence? How were the heavy rocks moved without modern equipment and why did they erect such massive buildings with huge columns — what was the purpose of the nearly identical pyramids in Egypt and in Central America? Maybe these civilizations knew something which was lost over the centuries? Maybe "modern" world religions WANTED this knowledge to get lost?

One additional, but most unusual discovery worth mentioning was made amongst the rocks and pillars off the islands of Bimini: A crystal ball with unusual mysterious powers was found, capable of deflecting magnetism and repelling certain metals. Also, the large figure of a fish etched in hardened sand under water was spotted first from low flying planes. Most of the discoveries were made by a Dr. Brown, who spent years to explore this area of many square miles, and a number of additional ruins are expected to be found buried deep below the ocean sands.

Speculation runs high amongst archaeologists that these discoveries may be a small part of the legendary civilization of Atlantis. Whatever the origin, it IS evidence that the Middle Eastern, Greek and Egyptian societies were not alone and may have been all part of ONE ancient (non-biblical) civilization.

Other discoveries have been made over the past 30 years, leading more and more to the acceptance of the kingdom of Atlantis as a historical fact, rather than a myth. The ancient Greek philosopher Plato wrote about it over 2000 years ago and we can find interesting parallels between the Minoan culture of Crete with Plato's Atlantis. The spectacular volcanic eruption of Santorini in 1470 B.C. levelled the Minoan island. As recent as 1966, remains of buildings have been discovered in the general area under about 250 feet of volcanic ashes and lava. The buildings were remarkable similar to those of Plato's description.

The psychic Edgar Cayce predicted in 1923 that the remains of an Atlantan (?) temple would be found under water in 1968 near the Bahamian Island of Bimini. In fact, ruins of a temple 75 ft. by 90 ft. WERE found in 1968 off the coast of Andros, also a Bahamian Island.

According to recent theories Atlantis may have been much larger than we ever expected. This speculation again is based on the under-water ruins near the Bimini islands, since they do display remarkable resemblance to other ruins in the Agaen Sea — thought to be from the time of Atlantis' existence. Also in line with Plato's notes is the absence of human remains amongst the building rubble and debris from an apparently advanced civilization. According to Plato, the inhabitants of Atlantis were warned (by extraterrestrials?) about an impending massive, volcanic disaster, which would wipe out all of their towns and dwellings.

As a legend, Atlantis goes back about 10,000 years and actual mysterious remains of giant statues of similar age can be found in various corners of our planet, thousands of miles apart. The objects could quite conceivably have been built by the same architects who designed and built Atlantis. The fact is, astronomical measuring devices and remains of great engineering works have been found dating back that far.

Whatever happened there, including the mysterious disappearance of the inhabitants of Atlantis, it makes the biblical story of Noah's ark look rather like a crude fairy tale. Particularly, when one accepts that Noah was to have taken two of each species into the ark — Earth had millions of animal species scattered all over the planet and below the water. As previously noted, they simply would not have fitted into an entire ARMADA of "arks." Another point to consider: Humanity has been around much longer and scattered over many more places than we ever learned from the Bible.

One of the loneliest places on Earth has to be Easter Island, occupied by Chile since 1888. The most recent population had been decimated by disease, war and slavery from over 10,000 natives to just over 1,500 souls. What is noticeable about the Island are over 500 mysterious huge statues carved into human faces. Their weight is anywhere from 30 to 80 tons each, and to this day scientists are unable to establish how they were moved and why they were built.

One theory is, the statues represent a number of "gods" (Aliens?) and when they failed to help the natives during natural disasters and battles with invaders, the locals purposely destroyed them during a symbolic "massacre." Another tragic example of blind "belief," which drove some of their people into death. Only this time, it happened without interference from western or eastern major religions. The story goes that just like the statues, their own people with big ears were massacred first. The most intriguing point seems to be the erection and the transport of the huge rocks. Who, when and why?

More startling evidence of very ancient, yet highly advanced civilizations, outside the biblical Middle East, is all around us. What appears to be an ancient observatory was discovered in the Yucatan, in a particular position which was obviously designed for tracking of the planet Venus at the time. A similar ancient astronomical observatory has been discovered in India.

The earliest known peace of cloth has been found in an old village in southeastern Turkey. The approximately 3 by 2 1/2 inch scrap was analyzed and found to be about 9,000 years old. Not too significant until one considers that in order to produce cloth, the locals must have had some sort of loom 7,000 years BC. Up to now, it was thought that the oldest woven cloth originated from a period between 6,500 and 6,000 years BC, but still long before the Bible was written and long before modern religions were introduced.

Also long before the Bible was written, the Tibetan "Book of Dzyan" records that in 9564 B.C. a large part of what is now the area of the Caribbean and the Gulf of Mexico sank into the ocean as a result of an unexplained cataclysmic event. How did the Tibetans know about 11,500 years ago what happened half-way around the world? Was it the end of Atlantis and did survivors tell them?

The Mayan civilization built astronomical observatories 3,000 years ago, and managed to chart the skies with great accuracy and the help of very primitive equipment, unaided and unobstructed by modern religions. They also created solar and lunar calendars with astonishing astronomical accuracy, and seemingly without known contact to other terrestrial civilizations.

Phoenician ruins can be seen at Mystery Hill in New Hampshire, having been carbon-dated at over 3,000 years of age. The architects apparently had a remarkable knowledge of astronomy, leaving us once again with the two obvious questions: first, why are the buildings so similar to those built 6,000 miles away? And second, why did this civilization know so much about astronomy without modern telescopes? The answers to both questions can not be found in the scripture. The existence of ancient ruins alone of approximately the same age and architecture, separated by thousands of miles is an obvious contradiction to "what is written."

From the late twelfth century until 1519, the Aztec civilization was at the zenith of their power with more knowledge about astronomy than their conquerors, the invading 500 or so members of Cortez's group of conquistadors. Unfortunately, the Aztec leader (Montezuma) did not fight what he thought to be the "people of the sun," until it was too late. The (good Christian) invaders massacred the majority of the Aztec natives, and forced the rest of them into slavery, later "introducing" Christianity to those who survived the massacre.

The kingdom of the Incas dominated the central Andes region from the 13th to the 16th century. Despite their talents in producing fine textiles, metal goods, massive stone buildings and roads, they did not use the wheel for transportation. The Inca civilization centered around Cuzco and Machu Picchu and at one time was to have numbered about twelve million people.

The question to this day remains: Why did they choose this totally isolated part of South America to build large cities? How did they transport tons of rocks to erect massive temples, from where they later worshipped the Sun? Machu Picchu is so far out of the way that it was only discovered in 1911. Maybe the Incas knew something we haven't learned from holy books?

Evidence was discovered recently that Buddhist monk explorers from China had landed about 578 A.D. at the coast of California, travelling at least as far as a so-called "GREAT Canyon" — presumably the "GRAND Canyon." At the same time, these visitors from China must have gone as far as Central America, where they visited the Mayan culture. Many Mayan statues of "gods" and serpents have an astonishing similarity with the same type of statues found in China of the same time period.

In California, a 48,000 year old human skull has been found, somewhat upsetting the old theory that native American Indians were the first settlers here. They were thought to have come from Asia via a previous land-bridge across the Bering Strait, about 20,000 years ago.

Of course, it is difficult to understand some of the less civilized traditions of the Mayans, the Aztecs and other ancient civilizations. An example of most savage Mayan "religious" rituals were offerings of sacrificial victims to the "gods" by these otherwise seemingly talented and advanced people. It is even more difficult to comprehend that the victims frequently volunteered to be sacrificed, while they were fully aware of the horrible way of dying. However, usually the victims were slaves or prisoners. Who instructed the ancient civilizations to sacrifice human lives to the sun "God," why and for what purpose?

THE AFRICAN DOGON TRIBE AND THEIR MOST INCREDIBLE LEGEND

Possibly the most remarkable human connection to extraterrestrials may be found within a small and peaceful African tribe in Mali. Maybe "Timbuktu" would be more familiar to the readers, it is located in the general area. Here, close to 250,000 natives live in relative harmony, in an area where they are not really included in any political structure and where they are not subjected to major world religions. Yet, common crimes such as murder or theft are almost unknown, and disagreements between tribal members are nearly always settled in a peaceful manner. That is only the beginning of the chapter on the "DOGON" TRIBE.

There are really four Dogon tribes — their ancestors divided the Dogon culture according to the compass, East, West, North and South. A number of Dogon villages are arranged with the huts and structures forming the figure of a man lying on his back. Again, this arrangement would be of no value to an observer on the ground. The origin of life is symbolized by the smallest cultivated seeds of the fonio grain, Kize-Uzi — "the little thing." Coincidental with my own observations about microorganisms as the real rulers of life, it describes the physical and social world of the Dogons as reality and it gives their "myths" a great amount of reasoning.

Most astonishingly, the majority of their legends are based on facts about the universe, some of which were only known to western civilization for the past fifty years. Yet, the Dogon myths go back to the dark ages and includes knowledge about the existence of rings around Saturn and of several moons around Jupiter, or about the Milky Way Galaxy. All of this knowledge could only be verified much later by telescope.

The ultimate ritual for any member of the Dogon tribe is based on the "dark star" as their source of life. During a 1937 tribal ceremony, Western observers were unable to explain the background for the very serious, yet festival atmosphere in honor of the "dark star." The ceremony is repeated about every sixty years as a once-in-a-lifetime event for most natives. Their orbital drawings of a set of twin stars, a "binary" star system, made no sense to the French visitors in 1937.

Meanwhile, astronomers have found the connection in more recent years when the binary system of Sirius A and Sirius B were discovered at over eight light-years distance from Earth. The bright Sirius A had always been known as the brightest star in the heavens for many centuries. Only mathematical calculation about its orbit and powerful telescopes actually revealed the existence of an invisible "twin." Its gravitational forces create a major influence on the orbit of the giant Sirius A. It was discovered that Sirius B is a "dead" star, a "White Dwarf," which may have died some time ago as a "Nova," the natural end of the life cycle for many stars. More about that and details on Sirius A and B in the chapters on astronomy.

The last time in 1937 and 60 years later in 1997, the Dogon tribe appears to celebrate the rare event of perfect alignment (conjunction) of Sirius A and B for the observer on Earth. How could a primitive African tribe possibly understand and have knowledge of this event hundreds of years ago, impossible to observe and taking place light years away from Earth?

Sirius A, the "dog star," still is the brightest star in the northern night sky and would be twenty eight times as bright as our sun, if it was to be placed in the same location as the Sun. It is therefore particularly remarkable that the physically much smaller, but extremely high density companion star, Sirius B, should have a major effect on the big "twin."

The great density of Sirius B is unimaginable at about 100,000 times that of water on Earth, causing the equally incomprehensible gravitational pull on Sirius A. However, the powerful gravitational forces of this white dwarf are quite normal. According to the Dogons' myth, messengers from this star system had visited them in ancient times and gave them knowledge about the universe and the origin of life. Considering the arrangement of their village buildings, one would suspect that the Dogons expect a return of the messengers from the Sirius star system, and most likely to happen during one of their 60-year rituals, which last several days.

How much more proof do doubters need for understanding and accepting the fact that we are NOT alone in the universe? That we are less technologically developed and much less "civilized" than some alien visitors from distant inter-galactic star systems? Unfortunately, even some astronomers tend to think the whole thing is just coincidence. Maybe these astronomers do not WANT to believe the facts of the Dogon legend, as it contradicts powerful biblical myth. As far as I am concerned, this fact-based "legend" provides more reality to humanity, than most biblical legends ever will.

The early Egyptian calendar was not based on the solar cycle, the motions of the star Sirius A were accurately applied instead. Is it coincidence that the Dogon tribe have also looked upon the Sirius pair of stars for hundreds of years as a source of extraterrestrial visitors? In both cases, why is Sirius so important? Do we find the answer in religious teachings?

One most revolutionary discovery was the so-called "Rhodes Calculator" in 1900 off the coast of the Greek Island Antikythera. At the time, it was thought to be a discarded nineteens century gadget. When it was found, it appeared to be some sort of clock mechanism, which had ended up in the sea along with 2000 year old debris from sunken Roman vessels. A British mathematician, Dr. Derek J. de Solla Price studied the mechanism about 58 years later and it was then discovered that it was indeed as old as the rest of the sunken "treasures." Not only was it built long before modern technology appeared on this planet, it was found to be a most accurate calendar based on the orbits of the sun, the moon AND the star Sirius! Since it was not built by humans 2000 years ago, the designers HAD TO BE extraterrestrials.

Gamma radiation usually penetrates everything, yet in an attempt to probe below the Khafre Pyramid in Egypt, archaeologists discovered a strange pattern of deflection from an applied field of gamma rays. Only another source of radiation deep below the pyramid would cause this behavior of the most powerful type of radiation. What is it and who put it there and why?

MORE CREDITABLE REPORTS OF UFO SIGHTINGS

Let us look at more "modern day" examples of reports about extra-terrestrial contacts.

A video of a UFO sighting was taken near Ottawa, Canada and it was thoroughly analyzed by a former NASA specialist and acknowledged to be genuine. It could not be explained as being of terrestrial origin.

Astronauts James McDevit and Gordon Cooper photographed UFO's from their spacecraft.

Two UFO's were photographed over a farm in McMinnville, Oregon in 1950. The photos were declared to be genuine, as well as the multiple eyewitness reports.

As the last entry on visits to North America by UFOs — an entire community observed the maneuvers of a UFO over the Hudson Valley between 1983 and 1987.

So far, I have usually concentrated on sightings over North America during recent years. UFOs have been observed by thousands of eyewitnesses from every corner of the planet. During the Geophysical year — over Brazil, in August 1968 over Rumania and Yugoslavia. March 1973 near Bangkok, Thailand. All these sightings were witnessed by a great number of observers. Again, the list would be endless, since very few countries did NOT report sightings.

To add it all up, the evidence of past AND present visits by extraterrestrial intelligent life on Earth is simply overwhelming. It is only a matter of time, when our governments will have to release previously hidden information to the public. All in all, we now have more "proof" of extraterrestrial life, than we will ever have about what is still taught in churches, synagogues and mosques.

It is most unfortunate that most factual reports and physical evidence have been kept top secret. Only occasionally do we have a "leak" in the system. One such leak was a report by General Nathan Twining about a number of what he called, "real objects of common general appearance," but of presumably extraterrestrial origin. Also, many individuals never report an encounter, or only after considerable time has elapsed after discussions with friends and family, since they fear ridicule. Particularly, if only one or two observers are involved. When hundreds of eyewitnesses observe the same occurrence, it becomes more difficult to keep it secret and the witnesses encourage each other to stick to their story about the encounter.

When several UFOs are seen by a number of air traffic controllers on radar screens in various locations, it is impossible to deny this event ever happened, or to discount it as a hoax. As noted earlier, our present day aircraft simply can not travel at the recorded speeds, nor can they change direction in an instant.

Once again, firm religious believers revert to ancient writings as the not-to-be-questioned "truth." When unexpected scientific facts, theories and discoveries do not agree with major religions, humanity is usually denied explanations about the unknown from the not-so-silent religious majority. At the very least, reports of UFO sightings are treated in a joking manner, or they are belittled and observers are persuaded into NOT believing what they saw. Nothing is ever taught in religious schools about it. Religions encourage questioning of scientific theories and facts, at the same time denying the right to question the existence of a superhuman "God" or "Creator". Only "evil sinners" would question holy teachings.

Little, if anything about alien intelligence agrees with the scripture, including the book of Genesis. According to it, one "God" created all life on Earth, humans and other creatures, about 4000 BC. Quite a difference from the actual approximately 550 million years of KNOWN existence of life on our planet. Yet, it was quite an acceptable theory at the time, considering the authors of the ancient holy books had little or no scientific knowledge.

To sum it up — sightings and encounters with alien visitors and spacecraft can no longer be ignored or written off, no more than the existence of civilized, intelligent life on Earth long BEFORE biblical times. The majority of reported UFO sightings may indeed be hoaxes or have a terrestrial origin. However, the incredible number of mysterious ancient structures of "alien" design, which can only to be seen from space, should be convincing enough. The similarity of these widely separated structures should eliminate all doubts about the reality of extraterrestrial visitations at one time or another.

One cannot ignore events such as the rituals by the Dogon tribe, their crude drawings which place Earth correctly in the Milky Way Galaxy, and their knowledge about the Sirius star system. How much more "proof" do sceptics need? Once again: where is the same overwhelming, modern day proof on biblical version of the creation of the universe and of humanity?

WHY ARE ALIEN LANDING SITES AND UFO SIGHTINGS USUALLY REPORTED FROM REMOTE AREAS?

One argument is frequently used by sceptics and nonbelievers in extraterrestrial life: Why are the aliens not landing in a more conspicuous place, such as a city square, where literally thousands could see them? Or better yet, where they could possibly communicate with a large group of observers about their intentions and their technology?

To answer that question, one must try to understand the minds of a highly advanced civilization and the reasoning for their visits to Earth in the first place. Based on the vast distances of interstellar space, it is obviously a very difficult task. Comparatively, the Judea-Christian-Islamic "God" has never landed in a city square either during the past 2000 years, at least not according to what religious teachings tell us.

Let us try to understand the reasons for alien visits. After all, WHY WOULD a highly advanced civilization want to explore this planet? It is obvious that their technology must be so far advanced and that mankind must appear to them as "stone-age" creatures. Their inter-stellar travel must take years, no matter how far the visitors are advanced and how fast they can move through space. Unless they have a large "mother ship" orbiting somewhere within the solar system, the living conditions for years in a relatively confined area of a thirty- to sixty-foot diameter "flying saucer" has to be most certainly less comfortable than life on their advanced home planet — unless their planet is dying.

I am eliminating the solar system as a likely source of alien visitors, since there are simply no planets or satellites nearby which could be reasonably hospitable for higher life forms. All planets and moons within this solar system are either too hot or too cold or have the wrong type of atmosphere. If, for example Venus had produced higher life forms in the past, they would have left a long, long time ago or the species would have become extinct. The same applies to Mars, where the current inhospitable conditions have prevailed for hundreds of thousands, if not millions of years. All other planets and their satellites are in worse shape and would not be suitable for even short-term residence. The reader can find more about that in the astronomy section.

If the intentions of alien visitors had been hostile towards mankind, with a planned take-over of the planet or possibly turning us into their slaves, they could have done so with great ease, hundreds or thousands of years ago. A fleet of twenty or thirty highly advanced UFOs, fully loaded with sophisticated weaponry could have done the job very easily, until the 1950's. After that, our own scientists had developed nuclear weapons and guided missiles to destroy any potential invaders.

That leaves one possible reason for their missions: a series of exploratory visits to all planets within this one and other solar systems. It would explain partly, why landings and sightings are usually reported only from remote areas of our planet, where the environment is still more "natural" and more unspoiled by modern man.

The visitors in all probability record, observe and occasionally take samples of soil etc., and take it all back to a mother ship or right to the home planet, along with samples and information from other planets from this and other solar systems.

Since some reports were of "encounters of a fourth kind," involving medical testing of abductees, it may also be possible that the aliens are involved in some sort of genetic engineering, trying to "transplant" some of the few advantages of the human body into their own kind. After all, the reported sightings of aliens describe them always as physically very frail (quite likely also a result of an evolutionary process), possibly resulting in physically weakening of their off-spring. A bit of genetic engineering with a still physically superior homo sapiens would improve that situation.

It is also important to remember that eyewitness reports do not necessarily describe all UFOs and their passengers as look-a-likes. One could therefore come to the conclusion that aliens do not originate from one and the same source, their individual home planets may be separated by light years and they may not even have contact with one another.

However, it appears that ALL of them are not too interested in communicating with us, since they obviously know everything we know and a lot more. They could not possibly learn from us. Human researchers study and experiment with ants and bees, but we have yet to try and communicate with them. Only recently have researchers studied communication between dolphins, chimpanzees and other higher life-forms.

On the other hand, when aliens are observing human behavior and the obsession to wage territorial and religious wars, they may not WANT to be seen too PUBLICLY and certainly would not want to announce a visit beforehand. Those crazy Earthlings may just try to capture or kill them with their nuclear weapons — just as they kill each other.

As far as interstellar travel time of several years is concerned, it is explained elsewhere. While alien spacecraft can probably travel at speeds near that of light, time for the occupants would pass much slower, at least if one of Einstein's theories is correct. While time on Earth and on their home planet would take eight or ten years for a return trip, the occupants of the interstellar spacecraft would only have aged a few months. Extraterrestrial visitors must have found a method to overcome the power of gravity or better yet, a way to use it to their advantage. They would therefore not have much trouble to travel great distances near the speed of light.

Since they are obviously "land animals" such as humans on Earth, a source of correct "air" mixture must be available during lengthy interstellar travel, along with food supplies. The idea of a mother-ship makes more sense all the time, since the supplies could be re-generated inside such a vessel. Another possibility would be travel of long distances under suspended animation (on autopilot?), some sort of hibernation.

Let me now go on to the most realistic substitute for religious myth, the science of "ASTRONOMY" and a more realistic source of "Creation" — the UNIVERSE, also known as the Cosmos.

SECTION II

THE TRUE SOURCE OF ALL CREATION — THE UNIVERSE

FACTS AND THEORIES OF ASTRONOMY MADE EASY AND COMPARED WITH RELIGIOUS TEACHINGS

Only a few hundred years ago it was not only considerably more difficult to learn and understand astronomy, it was also persecuted by religious teachers and leaders. The universe was taboo as it was said to be God's domain. Thomas Paine realized it 200 years ago and he would be delighted if he could talk to astronomers today. Here are his views as quoted from *The Age of Reason*:

"So late as 1610, Galileo, a Florentine, discovered and introduced the use of telescopes, and by applying them to observe the motions and appearances of the heavenly bodies afforded additional means for ascertaining the true structure of the universe. Instead of being esteemed for those discoveries, he was sentenced to renounce them, or the opinions resulting from them, as a damnable heresy. And, prior to that time, Virgilius was condemned to be burned for asserting antipodes, or in other words that the earth was a globe and habitable in every part where there was land."

Religious suppression of astronomy and the exploration of the cosmos in general has probably set mankind back by centuries until the recent introduction of modern technology. Everybody can now purchase a telescope and has the option to observe many of the mysteries of the universe.

During the past few decades, more than ever before, science has been re-shaping our concept of the cosmos. One of the difficulties in understanding astronomy is the usual manner of introduction to the layman, to students in public schools and junior high school grades. Textbooks on astronomy and most other sciences are written by highly educated science professors and are meant specifically for university post-graduates.

An individual with anything less than a university degree, including the average high school graduate and right down to young public school pupils, usually finds science textbooks too difficult to understand or "boring" and soon loses interest in this and other all important science subjects. Yet, all scientific knowledge which deals with life, its creation and the realities of the cosmos may serve as spiritual fulfillment to those who seriously question religious myth. That includes all confirmed and potential agnostics as well as all of the other over one billion "non-believers" on Earth.

To make humanity's oldest science, ASTRONOMY and the vastness of the universe easier to comprehend, the reader must first familiarize themselves with basic astronomical terms and try to comprehend interplanetary, interstellar and intergalactic dimensions. Let us start by standing (in reality or imaginary) on a pier at the shores of one of the Great Lakes in North America, and as much as possible away from interfering city lights.

Next, it is important for the observer to remember at all times, every visible object in the night sky appears as it WAS seconds, minutes, hours, days - but in the majority of cases — years ago — often hundreds and thousands of years. Looking at the night sky IS looking back in time, also known as "Time Dilation."

During a clear day or night, the observer may be fortunate enough to make out high structures on the other side of the lake. Example would be smoke stacks, transmission towers or very tall buildings, thirty, forty or more miles away. Noticeably, only the TOP HALF or so of a structure is visible, the first basic fact of astronomy that our planet Earth and almost all other heavenly objects are very large "globes."

With the exception of asteroids, all planets, our sun and all other stars are nearly perfectly round, contrary to ancient belief and teachings. Without man-made structures on the other side of the lake during ancient times, it was easy to believe that the Earth was flat. Even if there were mountains visible on "the other side," it would have been difficult to tell that only the upper portion of a mountain was visible.

If eyesight could follow the curvature of the Earth by travelling along the surface of our planet, first beyond the tower on the other side, and continuing all around the planet, an observer could see his own back. His sight would travel the circumference of Earth at about 24,000 miles, and at the speed of light in a matter of micro (μ = micro) seconds. Which brings us to the "speed of light," a basic part of all astronomical calculations.

A VARIETY OF RADIATION THROUGHOUT THE UNIVERSE

To understand the universe and astronomical terms in general, we must first try to comprehend the most important term in astronomy, the "speed of light" and that of radio and other waves. And why waves behaves the way they do.

Light is a narrow band of radiation, visible to the human eye. When we think of radiation, many readers may immediately think of X-rays or microwave radiation, both of them harmful to living cells. The big difference is in the FREQUENCY or "WAVELENGTH" of all radiation, light and otherwise. AM, FM radio and TV transmitters "radiate" at a lower frequency range than light sources, and unless one spends a lot of time extremely close to a high power radio transmission tower, radio frequency radiation will not harm any life-form.

In addition to the FREQUENCY of an emitting source, the POWER output is of greatest importance. Larger doses of power, even at lower frequencies can also be harmful to life. Light is a good example, when one looks into an ultra powerful source (sunlight, hi-power artificial light, arc-welding) for just a few seconds, the observer could end up with permanent damage to the retina of the eyes. Yet, looking into a low power, low frequency (60 Hertz in North America) light bulb is perfectly harmless.

Medium power infrared "radiation" sources (electric heaters) are also harmless to the eyes, since they emit at a lower frequency range than visible light. Before the eyesight would be damaged from getting too close to a high powered infrared radiation source, skin may be burned much easier by accidentally touching the heat source.

Beyond the highest visible light frequency "violet," the next higher (invisible) frequency is in the harmful ultraviolet range. Today, its damaging effects to unprotected skin and to plant life, even at lower power ranges, are known to everybody.

Continuing on into higher frequency ranges, such as microwave or radar, the danger to living cellular structures becomes more severe. Some of the most dangerous and highest frequency sources are X-rays and gamma rays with deadly effects, even at VERY LOW POWER output. As the "speed" of frequencies increases, and/or the power output, the damage to living cells increases accordingly.

The frequency of all radiation sources are measured in "*cycles*" or "*Hertz*" per second. Frequency is best understood as the number of waves passing each second, while "wave-length" is the distance between two peaks of waves. I will not use wavelength at this point, to avoid confusing the reader with too many different terms. Frequency measurements are also more easily understood.

The sound-wave frequency range is generally audible to the human ear, depending on a person's age, at between 16 Hertz and about 16,000 Hertz (16 kilo Hertz "kHz"). Certain animals can detect sound down to less than 10 Hz and others can "hear" frequencies in the 40 kHz range and above, as high as those of ultra-sound, (between 16,000 & 30,000 Hz, or 16 - 30 kHz.).

"LF" (low frequency) emissions of 30 to 300 kHz are

applied to little used low frequency radio bands. Medium-wave AM radio transmitters operate on a frequency range between about 450 kHz and 1,680 kHz, (or 1.68 mega Hertz, "mHz"). The VHF and UHF bands (FM radio and TV) operate within the 30 to 3,000 mega (million) Hertz (mHz) range, sometimes as high as 30,000 mHz.

Microwave transmitters and some other high-tech devices operate at frequencies above 3,000 mHz (3 giga Hertz), to 300 gHz and higher. Infrared radiation starts around 300 gHz. At this point, and to avoid super-high numbers, it will be simpler to understand the "speed" of emissions by expressing them as "wavelength" in "Angstrom units," A.U. Infrared radiation (heat), takes up to about 7,600 A.U. Visible light (red to violet range) comes in between about 7,600 A.U. and 3,900 A.U. (Note the *decrease* in *A. U.* as the *frequency increases.*)

When the radiation band between 3,900 to 320 A.U. is reached, we are already well into ultraviolet radiation and one of the most harmful frequency ranges. X-Ray emission ranges from 320 to 0.1 A.U., and the most harmful radiation of them all, gamma rays, start at about 0.1 A.U. up to 0.006 A.U., (an incredible 300 billion gHz frequency range).

As every high school student knows, X-ray radiation can be stopped by putting a shield of high density material such as lead between the "transmitter" and the operator of the equipment. However, the much higher frequency gamma rays in HIGH power doses will penetrate any ordinary lead shield or concrete of 6 to 9 ft. thickness. Only very controlled, minute doses of gamma radiation are actually used in industry for sterilization and in medicine in place of X-ray tests.

We are in reality constantly exposed to a minute dosage of gamma radiation from extraterrestrial sources, from the day we are born to the day we die. Gamma rays are able to penetrate the atmosphere much easier than ultraviolet radiation. Only permanent residence in a deep underground cave would protect us from this radiation. Fortunately, the intensity of gamma radiation penetrating the atmosphere from outer space is extremely minute and life on Earth has been conditioned to accept it through evolution.

Cosmic rays, at the highest known frequency range, (or lowest wavelength from any source) also bombard this planet constantly, equally influencing life from conception to death. As the name implies, this form of radiation originates from the cosmos, from distant sources near the centre of our galaxy. For the record: unless we are protected by a thick solid lead shield, or by at least 70 meters of water, nothing can escape the impact of cosmic radiation. It is thought, that natural aging of all living organisms may be caused partly by cosmic rays. And to close the "lesson" on radiation: Police radar operates usually on 10 gHz.

This rather basic and short lesson on radiation was absolutely essential for anyone with the slightest desire to understand astronomy. One final, but most important point to remember is the behavior of all radiation, from the lowest frequency radio band, to the highest giga Hertz range obtainable. ALL *DIRECT* RADIATION CAN BE SHUT OFF INSTANTLY when the source is disconnected. Just as a light switch, holding a black shield in front of the sun's rays, or the switch of a microwave oven disconnects the source in an instant. Or a wall of seventy meters of water will shut off even cosmic radiation. However, only a complete enclosure will protect from "stray" or "reflected" radiation.

Ultra high frequency sources, such as X-ray equipment have to be shielded by lead before the equipment is activated, yet a flick of a switch will also disconnect the transmitter. Therefore, there is no harm in putting one's hand into a microwave oven immediately AFTER the unit is shut off. For nearly complete protection from Gamma and Cosmic rays and to allow uninhibited research, shelters are actually built with concrete walls of several yards' thickness.

Finally, for practical purposes, all radiation travels in a relatively *straight line* and can be deflected in all directions. A good example is the deflection of light by mirrors. With the exception of sound, all other waves travel at the same incredible speed — the "SPEED OF LIGHT" — 186,000 miles per second, (about 300,000 km per second).

Although visible light travels in a "relatively straight line," there is an exception. According to one of Einstein's theories, light and radiation does NOT really travel in a straight line over the great distances in interstellar space. Light and radio waves are not only reflected by interstellar objects, the beams travel with a calculable curvature. This effect is so minute that it is noticeable only over the vast distances in the universe, with some light rays travelling billions of light years before reaching us.

Visible light and other radiation is also DEFLECTED by various objects in space. The deflection may be as minute as the CURVATURE of light beams over distance, but the great number of objects on the way WILL have an effect. While passing by millions of stars, star clusters and planets, the shift becomes noticeable enough to alter the apparent location of distant galaxies. This factor takes into consideration Einstein's theory of "curved space" whereby space itself is curved. Adding up all the various effects of deflection and curvature of light rays, it becomes possibly somewhat difficult to comprehend for the layman. But for all practical purposes, I will use "line of sight," straight travel for light in the following chapters.

"Curved space" also translates into a strange reality, theoretically allowing us to "see" our own backs, if we COULD see anything at all at the "other end" of the universe. Of course this theory is impossible to apply, since we would observe the space where we stand now, how it was as much as fifteen billion years ago. But then, many of Einstein's theories HAVE already been proven, this one may be next in one way or another. He may have been THE ultimate "prophet."

Einstein's theory of light deflection by other interstellar objects HAS actually been proven by twentieth century science. Proving the theory of curved space and curved light rays may explain another theory, the one of an infinite Universe.

THE SOLAR SYSTEM, PLANETS, SATELLITES AND ASTEROIDS

Let us go back to Earth, back to our observation point at the pier at one of the Great Lakes, away from city lights. Once again, we can plainly see that our planet is a globe, just as the Moon or the Sun. It is difficult to understand under the weight of obvious evidence that a large group of intelligent people believed during recent centuries in a "flat" Earth. Maybe, they accepted as fact illustrations found in old editions of the Bible?

The organization called "Flat Earth Society" really has never completely disappeared. It is known to astronomers that ALL planets, just like the stars, were in a gaseous or liquid state at one time. Any "liquid" matter in the near-vacuum of space will behave in the same manner, as long as there is no major gravitational pull from another source nearby. A liquid will always form a near-perfect "ball" under its own gravitational force. As the planets cooled off and hardened, their original shape was simply retained and all lighter liquid or gaseous matter (water and the atmosphere on Earth), is held captive to the surface by our planet's gravitational forces.

Each larger body of greater mass applies a stronger gravitational force upon a smaller nearby object, it attracts lighter bodies. This effect can be demonstrated in a calm sea, when a small boat comes too close to a large ocean liner and it seems to magically drift towards the large vessel. Because of the low resistance from water to the movement of the small boat, both objects over-ride the much more powerful effect of the Earth's gravitational field. As the core of our planet is of much greater mass than water, the oceans are held into place at approximately the same level, at "zero sea level." Only the moon's gravitational pull alters the normal sea level somewhat, causing the daily tides.

Let us now look at the evening sky. It is a clear night and the half moon is a pleasant sight. Since it is about

240,000 miles away from Earth in a straight line of sight, the light we see has travelled over one second at 186,000 miles per second to reach us. Once again, the larger body attracts the smaller object, as Earth with a diameter of nearly 8,000 miles attracts the moon of only a bit over 2,100 miles diameter. Fortunately, this action is precisely counteracted by the orbital speed of both objects and the distance stays practically unchanged. It has been going on for nearly 5 billion years.

Had the Moon been moving slightly slower in its orbital motion or had it been closer to Earth billions of years ago, it would have either collided with our planet a long time ago, or a much tighter orbit would have eventually had similar devastating effects on Earth with the moon breaking up in pieces before collision. Astronomers had an excellent example of this phenomena when the Shoemaker-Levy comet broke up and the chunks of it collided with Jupiter.

Astronomers have discovered large craters not only on the Moon, but also on some planets in the solar system, such as Mercury, Mars and Venus. When the planets and satellites in the solar system were still in a semi-solid state shortly after it had formed, many smaller bodies were captured by their larger "neighbor's" gravitational pull. A few billion years ago, it must have been quite a mess out there.

Early inter-planetary collisions have altered the orbit of some of the "capturing" bodies after impact, leaving huge visible craters on the surfaces. Erosion through weather on Earth has erased most of the craters, only some of the more recent collisions are still traceable, such as the ancient Meteor Crater of Arizona. The point is, there is no difference in the behavior of one "heavenly" body compared another. Only mass, orbital speed, diameter and age have their precisely predictable and calculable effect on Earth and all other planets in the solar system, all according to the laws of physics.

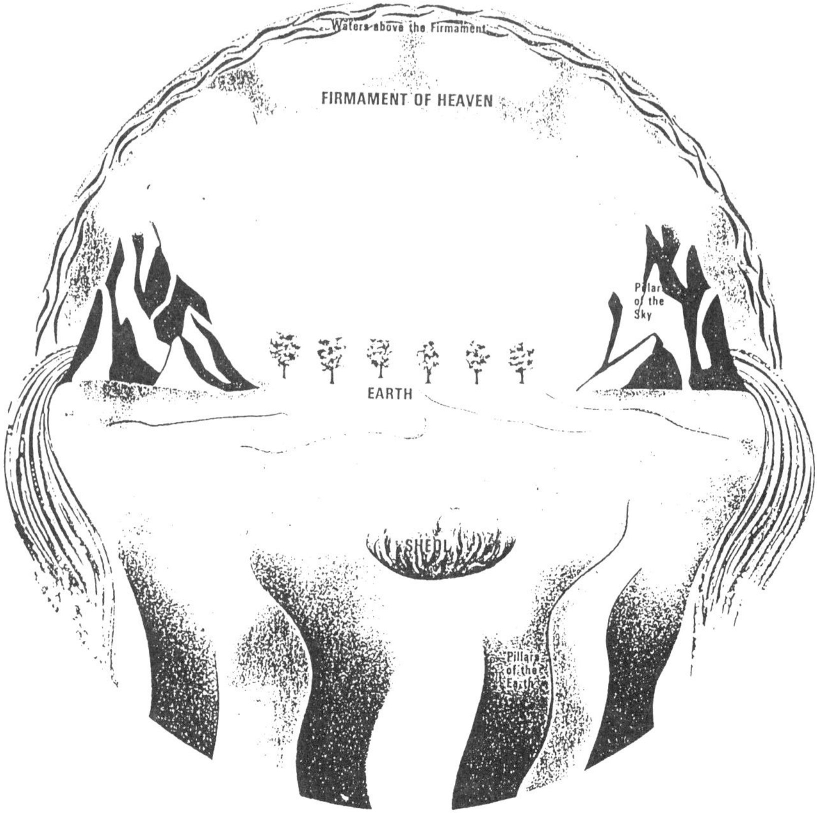

Illustrated: The Ancient Hebrews and Mesopotamians had no alternative, they HAD to believe (or accept as FACT) this image of the Universe. Today, why are so many believers still not accepting the scientific facts of the Universe? If it is written (or illustrated) in the Bible, it must be undisputable "Truth"?

THE PLANET MERCURY NEVER REALLY HAD A CHANCE

At only about 3,000 diameter the planet MERCURY has a lot less gravitational pull than Earth, and it was therefore unable to hold water and an atmosphere, if there ever was any. Our satellite, the Moon, was subjected to the same conditions as Mercury due to their similarity in physical dimensions. Because of their small size *both* objects were unable to retain liquid surface water and atmospheric gases. The effect from the lack of water and atmosphere resulted in virtually "no weather" and no change to the surface through erosion, contrary to Earth. Without water, even rather simple life-forms obviously cannot exist on Mercury or on the Moon now or in the recent past.

There may have been a brief period, when PRIMITIVE life forms COULD theoretically have developed on Mercury or the Moon, but because of the prevailing conditions, they could not have survived for long. Single-cell life would have been unable to develop further through the evolutionary process. To make things worse for Mercury, it is orbiting the Sun at only about 36,000,000 miles distance, much too close for comfort. Because of the tremendous gravitational pull from the nearby sun, Mercury's ORBITAL speed is at an incredible 30 miles per second, thereby completing a Mercurian year (one complete orbit around the sun) in 88 Earth days. While Earth rotates on its axis once in 24 hours, Mercury needs nearly 59 days to complete one day and one night.

Almost 30 days of unobstructed, scorching hot sunshine without a protective atmosphere causes "day-time" high temperatures of up to 430° C on Mercury, while during the long night it is estimated to drop to a more comfortable near "room temperature" of about 20° C. However, other scientists are of the opinion that the night temperature may be as low as minus 180° C. The difference of opinion is a result of the uncertainty about the presence of even a most shallow atmosphere of possibly carbon dioxide, which would help to retain some of the solar heat during the long Mercurian night.

Getting back to our observation point at the pier, with the aid of binoculars one can actually see the huge craters on the Moon, particularly near the inner edge of a "half" moon. However, since the planet Mercury is nearly 60,000,000 miles further from Earth than the moon and it is relatively close to the Sun, we can only observe it occasionally shortly before sunrise or immediately after sunset. Craters on Mercury are not visible through binoculars, only through professional telescopes.

Comparing unmarked photographs of Mercury and the Moon side by side, provides excellent proof of how similar "heavenly bodies" of similar size appear to the observer. In fact, anyone who is not familiar with some of the surface features of the Moon may easily mistake one for the other.

VENUS, OUR INHOSPITABLE "TWIN" — OR THE BIBLICAL "HELL"?

Earth's "twin" planet VENUS could have been more suitable for life of some sort to develop millions of years ago. Its diameter of about 7,000 miles is similar to the Earth and could hold an atmosphere and surface water. The distance of Venus from the solar "nuclear oven" is considerably more favorable than for Mercury, at about 67,300,000 miles from the Sun. From our observation point, we can see Venus unmistakably as the brightest "object" in the night sky. At its closest orbital position it is "only" about 25,000,000 miles from Earth. What we see, took 135 seconds to reach us, over two minutes. An incredible difference in distance compared to the Moon! And that is the planet CLOSEST to Earth.

Although Venus is similar to Earth in size and it is the nearest planet to us, much is entirely different from Earth. The Magellan spacecraft's mission came to an end in late 1994, giving astronomers a greatly "up-dated" perspective on the neighbor planet. It has now been confirmed that a rather dramatic occurrence appears to have re-shaped the entire surface of Venus between 300 and 500 million years ago.

Not connected with this last mission are scientists' earlier observations that Venus as the only planet in the solar system spins CLOCKWISE. All others rotate counterclockwise. One theory identifies Venus as an intruding object, which ended up in its current solar orbit after coming from interstellar space about 500 million years ago. Coincidentally, Earth had undergone dramatic changes also about the same time with drastic shifts in the surface and alterations to the atmosphere. And strangely enough, the first primitive life on Earth also dates back to this period. The upheaval to the orbital motions by the intrusion of Venus so relatively close to our planet may have had a lot to do with it.

If one could observe Earth from a vantage point on Venus, our planet would appear about the same size as Venus does from here. Earth would be the brightest object in the sky, aside from the Sun. Venus to the terrestrial observer appears very bright due to a heavy cloud cover, which reflects sunshine back into space. The permanent cloud cover also acts as a green house, keeping the searing surface heat from escaping. The surface temperature reaches 475° C, enough to melt lead.

To make the planet even more inhospitable, the heavy cloud cover seems to consist partly of sulfuric acid, and mostly of carbon dioxide. Only a tiny trace of water vapor has been detected so far. Surface winds of up to 200 miles per hour keep the heat evenly distributed, including on the night side where the temperature only drops a few degrees during the 58 Earth- "day" -long night. A very hostile place indeed.

Maybe religious believers should consider Venus as the ultimate biblical "Hell"? But then, the Earth's interior is just as hot a hell, as some of those of my friends tell me, who did not make it to heaven "up there!" But are we not ALL going to turn from "ashes to ashes?" Very confusing, I would rather stick to scientific facts.

The "Pioneer" Venus probe, before burning up in the Venusian atmosphere in the fall of 1992 transmitted some very startling evidence about the existence of an ancient ocean on Venus. The body of water may have existed for a period of about one billion years, enough time for primitive life-forms to develop. Chemical residue left behind is conclusive evidence for the existence of an "ocean" of perhaps 25 meters (about 70 ft.) depth. Scientists have reason to believe that the Sun's emission of heat and light in general was about 30% lower several billion years ago and it could therefore have been cool enough for water to remain on the surface of Venus for nearly one billion years.

Later, due to increasing heat from the Sun, carbon dioxide and the existing water vapor itself ultimately caused a severe greenhouse effect in the atmosphere. The surface heated up enough to eventually trigger the disappearance of surface water and vapor. However, other scientists are more sceptical about these findings. The fact remains, there HAS BEEN water on Venus some time ago, the question is how much, in what form (surface water, vapor or both) and how long was it present? This discovery also proves one point: all chemical elements behave in the same manner on all other "heavenly bodies," EARTH IS NOT AT ALL UNIQUE.

MARS, OFTEN THOUGHT TO BE THE CLOSEST "RELATIVE" TO EARTH

Depending on time of the year, we may be able to see the planet Mars from our observation point as a tiny reddish dot in the sky. Depending on orbital location, this planet's distance from Earth averages around about 50,000,000 miles. Due to the planet's eccentric orbit around the Sun, the distance from Earth really varies greatly; to be exact, from 35,000,000 to over 63,000,000 miles.

At approximately 268 light seconds (mean) distance from Earth, reflected sunlight from Mars takes nearly 4½ minutes to reach the observer at the pier. Compared to Earth, solar radiation received on the smaller martian surface is much weaker due to the greater distance from the Sun at an average 141,500,000 miles. In addition, the planet cannot hold as much of an atmosphere as Earth or Venus based on lesser gravitational forces of the smaller Mars. Its diameter is only about 4,200 miles, about half that of Earth.

However, scientist still believe that a martian atmosphere did exist for a "brief" period of at least a few 100,000,000 years or so, enough time for primitive life to develop on Mars and to adapt perhaps even to today's hostile martian environment by living underground. Mars has very sparse remains of frozen surface water at the present time, but it appears from visible canyons that flowing rivers may have existed. Also, the south pole region shows signs of earlier glaciers.

The planet appears to have no atmosphere as we know it, only a very "thin" mixture of about 95% carbon dioxide, 0.2% oxygen, 2-3% molecular nitrogen and one to two percent argon. The atmospheric pressure at the surface has been measured to be a mere one to two percent of the Earth's surface pressure, or about equal to atmospheric pressure at 25 miles altitude on Earth. However, Mars could support surviving primitive life from "better" days, with enough time for evolutionary adaptation.

So far, scientists have not found any evidence of present or past life on Mars. The current lack of an appreciable amount of surface water and the sparseness of an atmosphere simple is too inhospitable for HIGHER life forms to have developed in the past. However, it is quite conceivable that LOWER life-forms may really have existed on Mars countless millennia ago when water was somewhat more abundant on its surface. After all, where there is water on Earth, there is life.

Life on Earth has been in a more fortunate position

as far as the supply of water is concerned, compared to Mars and to Venus. Of all three planets, only Earth has retained the surface water to this day, all the way through a rather parallel evolutionary process. Mars apparently had at one (short) time retained an even higher percentage of surface water than Earth. Mars simply lost its water through much more rapid evaporation, due to its much lower surface gravity.

Venus on the other hand lost its water due to excessive heat, in comparison. The early "steam" clouds in Venus' atmosphere were rapidly split into the basic atomic components of water: Hydrogen and oxygen, aided by exposure to ultraviolet solar radiation which is much greater than on Earth.

The lack of a protective ozone layer allowed the process of escape to continue for both gases (particularly hydrogen), from Mars and Venus. Without doubt, a considerable volume of water also has evaporated into space from Earth, but at a much slower pace. Evaporation will increase for all three planets when the Sun comes to the end of her days and becomes a "red giant," four to five billion years hence. More about that in a later chapter.

Mars travels at about 15 miles per second in its solar orbit to complete one Martian year of 687 Earth days. The Martian day has about the same duration as that on Earth, about 24½ hours. And finally, startling recent discoveries revealed unexpected volcanic activities on Mars, much more powerful than those on Earth. The surface temperature on Mars is not inviting at well below zero most of the time, or at best around freezing.

In orbit around Mars are two "tiny" moons of irregular shape, Phobos at about 8 by 12 miles "diameter" and Deimos measuring about 5 by 7½ miles. Before astronomers were able to obtain photographs from both Martian moons, laymen and astronomers both suspected Phobos to be either an artificial satellite or at least to have been altered into an artificial observation point for intelligent extraterrestrials.

Phobos displays some unusual features: It rises in the WESTERN horizon of Mars and sets in the EAST. Phobos is of pronounced elliptical shape and tumbles in orbit around Mars with three axes. The features of Deimos are similar, but perhaps not as noticeable and this tiny satellite also behaves more "normal" by rising in the east and setting in the west.

Strangely enough, both moons were already known to the ancient Greeks, long before discovery was possible through powerful telescopes. They were first observed by the American astronomer Asaph Hall in 1877. Another astronomer, Johannes Kepler of Germany suspected their existence as far back as 1610. Jonathan Swift described both moons with great accuracy in part III of Gulliver's Travels, long before they could actually be observed by today's astronomers.

To the observer here, Mars and all other planets of GREATER distance than Earth from the Sun, will appear in a "full" phase, like a full moon, since we are on the "inside" looking "out." Mars is an exception, at times it can be in an orbital location which makes it appear crescent-shaped through a telescope.

To explain "inside looking out" a bit further: When the outer planets are visible in the night sky, the terrestrial observer has the sun shining from "behind," thus creating the permanent "full" phase image.

ASTEROIDS, MYSTERIOUS AND DANGEROUS COMPANIONS IN THE SOLAR SYSTEM

Before we look for the next planet — the easily recognizable Jupiter — we must not forget the ring of ASTEROIDS. They are basically orbiting between Mars and Jupiter and usually invisible to terrestrial observers, except through the most powerful telescopes. Only one asteroid, Vesta, is faintly visible at times to the unaided eye. Vesta is the third largest asteroid at about 310 miles diameter.

The second largest "micro planet" is Pallas at about 340 miles diameter, while the largest of about 2,000 known asteroids is Ceres at about 625 miles diameter. It orbits around the Sun once every 4.6 years. Since it consists mainly of dark rock formations, it is not at all visible to the unaided eye.

Until recent years, asteroids were thought to be the remnants of a former planet, left over from a major collision of planets. The evidence now points to their origin as leftover parts from the time of formation of all existing planets. They simply have not yet orbited close enough to a nearby planet, disintegrating and merging with one of them, usually with Jupiter or Mars.

However, it is quite conceivable for collisions to occur on a regular basis, as some of the smaller asteroids stray out of their orbit frequently, getting trapped in the orbit of a planet. The original theory of two colliding small planets during the formation of the solar system, is still more acceptable as a birthplace for asteroids to some astronomers. It is difficult to discredit one theory over the other, since both are based on logical conclusions.

These "planetoids" are of no significance to us until one of them strays out of its orbit and gets into the Earth's gravitational forces. Depending on the counteracting forces of terrestrial gravity, distance, speed and size of the asteroid, such an event in the distant past can cause and may have resulted in the worst possible natural catastrophes. Worse than all past hurricanes, earthquakes and volcanic eruptions combined.

As noted earlier, collisions of large meteors or asteroids with Earth indeed had devastating effects to all life on this planet. Many scientists believe that one such event caused the extinction of all large creatures on the Earth's surface, mainly the dinosaurs about 65,000,000 years ago. An asteroid would have to be of a diameter of the best part of a mile or bigger to cause major, long ranging changes of this magnitude to the Earth's surface and to our atmosphere.

Scientists today have mounting evidence of this or another catastrophic event from sediment samples taken in the "Chicxulub" crater beneath the Yucatan Peninsula in Mexico. Early indication were, this crater has a diameter of about 300 km (190 miles) and it was in all likelihood caused by a comet, rather than an asteroid. Whatever it was, the results had to be devastating to the entire planet about 65 million years ago. "Tektite" glass fragments, (solidified droplets of briefly molten rock) have been found, originating from this impact. This crater is one of the largest anywhere to be found in the solar system.

Dust and debris in the atmosphere, resulting from the massive impact, would have caused sunlight to be almost totally blocked out for several months or even years. Resulting darkness and rapidly dropping temperatures would unquestionably have caused the extinction of most plant life, the only source of food for many species. Large plant-eating animals with an appetite like that of dinosaurs would be the first to die. After the initial cold spell and darkness, remaining fine dust particles in the atmosphere would have caused the opposite (greenhouse) effect once the sun was able to penetrate, resulting in a heat wave for hundreds of years, perhaps a thousand years.

The impact would also have caused massive floods, resulting in unimaginable destruction to the entire planet, (maybe the "biblical flood"? However, 65 million years ago, there were no punishable "sinners" to be found on Earth and according to religious teachings, the Earth and the Universe did not even exist 65 million years ago!

The resulting turmoil from an impact of a meteor or an asteroid, possibly no larger than 10 to 20 miles in diameter, would have to be simply awesome. Scientists estimate the devastation from a meteor or asteroid of "only" 10 km (6 miles) diameter crashing into Earth at about 25 km per second, would be comparable to the explosion of almost ALL COMBINED nuclear weapons of the Unites States and Russia. How much more devastating would an impact be on Earth from a hypothetical 400-mile-diameter asteroid? However, considering if and WHEN a collision of this magnitude had happened, erosion would have wiped out most traces by now.

Some scientists believe, the Pacific Ocean immediately west of Japan, Taiwan and the Philippines could have been a very likely crash site for an extraterrestrial object of truly giant dimensions, billions of years ago. Looking at a map of the area, and considering that the Pacific Ocean is at its deepest at that general location, the makings for a major asteroid crash site during the early days of Earth are rather apparent and could be a logical and fact-based theory.

Major impacts of asteroids or comets onto Earth over the entire history of the planet are simple, explainable acts of physics and chance. Scientists can forecast repeat performances today through observations and calculations, warning us if we were to fall victim again to this type of unimaginable devastation. As mentioned, the legendary biblical flood could have been caused by a more recent asteroid impact, or a near-collision causing changes to the Earth's orbital patterns.

Ceres, the largest asteroid, is nothing really to worry about for earthlings, since the mean distance of Ceres from the Sun is about 257,000,000 miles. Our planet is orbiting at about 93,000,000 miles from the Sun, the difference in distance therefore puts Ceres far enough away from here. We DID have a few close "encounters" with asteroids in recent past: Apollo approached Earth's orbit within 3,000,000 miles in 1932 — Adonis came within 1,000,000 miles in 1936 — Hermes came really dangerously close in 1937, at only about 400,000 miles distance, and Eros came within 19,000,000 miles of the Earth's orbit in 1931.

Compared to the Earth's orbit around the Sun of 365¼ days, most asteroids take between 3,200 and 5,500 years for one orbit "one asteroid year," due to their greater distance from the Sun and small dimensions.

I have given more importance than readers might expect to the ring of several thousands of asteroids, compared to other planets and satellites orbiting our Sun. One collision with a single asteroid will change the entire Earth environment. If anything, no other planet can have, or had as much influence on planet Earth. Scientists are still investigating the effects on Earth and on evolution, from possible past collisions with asteroids.

JUPITER, OUR GIANT COMPANION IN THE SOLAR SYSTEM

Now it is time to search from our pier for the next "closest" planet, JUPITER, the largest of them all within the solar system. Its polar diameter is 82,800 miles and the equatorial diameter is considerably greater at 88,700 miles. Just for a quick comparison, the Earth's diameter is under 8,000 miles.

The considerable flattened polar regions are a result of the planet's rapid rotation of less than ten hours (one Jupiter day) and the relatively low density of its mass at only one quarter that of Earth. Still, as the largest planet in the solar system, it combines greater mass and volume than ALL OTHER planets together, 1403 times Earth's volume — 318 times the mass. It makes our own little planet look very insignificant indeed!

The fast spinning action of Jupiter causes the "atmosphere" and the surface to be forced outwards at the equator against the relatively low gravitational pull from the core. Jupiter's great volume and gravity of 2.64 times that of Earth's has prevented nearly all low density matter from escaping, including most of the light hydrogen gas, thus making it a "gaseous" planet.

Conflicting numbers give the mass of Jupiter between 2½ times and over three times that of all other planets combined; in any case, a 150 pound person would weigh between 375 and 450 pounds on Jupiter. The gravitational force of the planet's greater mass held the lighter gases captive contrary to Earth, where hydrogen and helium gases are dissipating fast into space. As a logical result of it, Jupiter's atmosphere is quite dissimilar from Earth's of about 99% nitrogen and oxygen. Jupiter retained nearly that much hydrogen (96%) in the atmosphere and in up to 12,500 mile deep "seas" of liquid hydrogen. Other gases found are mainly methane and ammonia.

Jupiter's orbit around the Sun takes about twelve Earth years (11.86 years to be exact), giving it six years of "summer" and six years of "winter," if we can call it that. Since Jupiter is the largest planet in the solar system, it is easily visible from our vantage point at the pier, despite its vast distance from Earth. That is of course, if it is visible above the horizon at any given point and time. Jupiter's mean *distance from the sun* is 484,000,000 miles, or when in line (conjunction) with Earth, about 391,000,000 miles from our planet.

When we observe Jupiter, it appears as it was about 35 minutes ago, provided Jupiter's orbital location is not at its maximum distance from Earth, when it takes even longer. Sunshine travels "only" eight minutes to reach Earth, in comparison.

My main objective in the astronomy chapters is to introduce the reader to a better comprehension of the vast distances and dimensions in the Universe. My effort goes beyond just supplying statistical numbers, the emphasis is on how insignificant our own planet is compared to the rest of the universe. Jupiter is a good example of the vastness of just the solar system alone.

Jupiter has managed to capture at least seventeen satellites (moons) and orbiting around the planet at distances between 260,000 miles (about the same as our Moon from Earth) and as much as 1,170,000 miles.

The four largest moons are visible through binoculars. To name a few of them: Ganymede at 3250 miles diameter is also the largest satellite in the solar system and has a diameter greater than the planet Mercury

(3100 miles). Callisto's diameter is about 3000 miles, while Io at 2250 miles diameter is also closest of the larger moons to Jupiter and displays the highest volcanic activity anywhere in the solar system. Possibly *because* of its closeness to Jupiter and the resulting high orbital speed of 42½ days. This orbital speed translates into 21 Days of "summer" and 21 days of "winter," if we want to apply a comparison.

"Europa" is the smallest of the four major satellites at about 1940 miles diameter, but it has an interesting surface. It is almost completely covered by ice, and possibly next to Earth, the only object within the solar system to hold a significant amount of water in one form or another.

Once again my point is, Earth is not at all unusual or unique, as one may believe from reading holy books. It is suspected that Jupiter's powerful gravitational pull captured some of its smallest satellites from stray asteroids, considering the unusual orbit of several of them.

Temperature measurements on Jupiter indicated a large variation from about minus 100° Fahrenheit (- 70° C) to a pleasant 80° F, taken at an atmospheric pressure of about one bar (equivalent to sea level atmospheric pressure on Earth). After penetrating to a depth of about 400 km (250 miles) below the one bar level, the atmospheric temperature increases to around 730° C (or 1,340° F), within a very inhospitable layer of heat. And that is not even surface temperature.

Jupiter emits more energy than it receives, and it is highly probable that fusion reaction still takes place within the dense core of the planet. Some astronomers consider Jupiter to be on the verge of a cross between a star and a planet, since it is displaying features of either a stellar body or a planet. However, there is no reason to believe that Jupiter is unique within the galaxy in that respect. The closest star to the solar system, Proxima Centauri is similar to Jupiter in that respect.

It is difficult to detect a surface on Jupiter as such. Under the "surface" pressure caused by the higher gravitational force, some gases may have turned into "slush" layers or into "slurry." Our space-probes, "Pioneers 10 and 11" provided confirming evidence that most of Jupiter's core is almost entirely made up from hydrogen. A "tiny" solid core of the size of Earth only is surrounded by liquid hydrogen and a molten "metal-type" mixture of hydrogen and helium at high temperatures and under tremendous pressure.

As an interesting statistic, in order to reach the vicinity of Jupiter, a man-made space probe has to accelerate to at least 39 miles per second.

Since our next and last clearly visible planet will be SATURN, I want to refer to the composition and temperatures of the atmosphere of Saturn at this time also. The conditions appear to be very similar between Jupiter and Saturn, for all practical purposes.

At the end of the chapter on Jupiter, believers and non-believers may ask themselves the question: why did "God" "create" these hostile planets surrounding us, just to have "created" one tiny planet reasonably hospitable for a creature of his image? For the reader who tries to compare the scientific, mostly predictable and understandable creation of stars and solar systems with ancient holy books — scientific FACTS are a much more reliable, acceptable and "believable" choice.

Religious legends do not teach us any of the facts of the universe, scientists had to find the answers and the proof to the theories without help from past biblical "prophets" or present day theologians. The Scripture does not tell us what happens to hydrogen under various pressure and temperature extremes, it also does not tell us, that at about 92% of all matter in the universe, or that the most predominant and "mother" of all elements — is HYDROGEN.

SATURN, THE SECOND GIANT AND A MAJESTIC SIGHT

From our pier at the edge of the lake, we can now search for the planet SATURN in the night sky. Just like

Venus, Mars and Jupiter, it is easily visible, when it is above the horizon. At a mean distance of about 886,000,000 miles from our Sun, the distance is almost twice that of Jupiter. When we see Saturn, the light has taken about 70 minutes to reach us.

To be precise, keeping in mind that all planets only reflect sun-light, I have given the amount of time elapsed only for reflected light to reach us from the planet. We can *double* this time period for light from the Sun to reach a planet under observation, since sunlight is "mirrored" only by the planets. After most of it is reflected back into space, the observer on Earth can still see enough light to identify a planet. To simplify it, when shining a light into a mirror, the light beams have to travel from the source to the mirror and back to the observer, to be seen.

Saturn is only slightly smaller than Jupiter, at a diameter of about 71,500 miles. Its mass, density and powerful gravitational pull has managed to keep at least 22 moons as captives, at distances between 116,000 and 8,000,000 miles. The largest, Titan, has a diameter of about 3000 miles, again almost as big as the planet Mercury. The exact number of satellites is not yet clear, since some of them appear to originate from the rings or they orbit within them.

INTERESTING DISCOVERIES IN THE ATMOSPHERE OF SATURN'S MOON "TITAN"

One particular satellite of Saturn is of greatest importance to scientists, the giant moon TITAN. It was one of the most significance discoveries to astronomers and to evolutionist, when Voyager 1 instruments discovered organic, non-living material in the satellite's atmosphere. Until then, organic molecules were thought to be unique to Earth, with the exception of earlier discoveries of amino acids and other building blocks of life in primitive meteorites.

However, the earlier detection of elements necessary for life in the remains of earth-bound meteorites were questionable, since the intruding object may have been contaminated by earthly elements, prior to or after crashing onto the surface.

An artificial atmosphere was created in recent years in a laboratory experiment, by combining the right mixture of methane gas, ammonia gas and water and an electrical charge (artificial lightning). The result was as expected, amino acids and other basic building materials were created — the fundamental essentials for the creation of life. Saturn's moon Titan has nearly exactly this type of atmosphere, necessary for the recipe to make life's primordial "soup."

The space probe "Voyager 1" recorded a rather dense atmosphere on Titan with a surface pressure 60% greater than on Earth. Thick rain clouds of mostly methane contained organic molecules and an atmosphere of nitrogen and carbon-based molecules may just have all the right ingredients for the creation of life. However, Titan's vast distance from the Sun and its extremely low surface temperature of approximately minus 290° F may have never allowed life to actually develop any further — but the ingredients are certainly there!

A European-built space probe will be launched in 1997, and if successful, the "Cassini" mission will provide scientists with further detailed information about Saturn and Titan. The "Huygens" probe is designed to descend through Titan's atmosphere, take a variety of measurements of the atmosphere and hopefully land on the frigid moon. The surface is thought to hold lakes and oceans of liquid ethane and methane and the bottom of the lakes and the land surface may be covered with complex carbon compounds. Although launch is scheduled to be in October 1997, the Cassini/Huygens mission will not reach its destination until 2004, travelling at incredible speeds between around 25,000 and 30,000 m.p.h.

Titan appears very inhospitable for the creation of life.

Considering the extreme cold, it is difficult to understand that molecules essential to life have survived. But then, life exists in the most unlikely places on Earth and the same molecules were here also, when the terrestrial atmosphere and temperatures were most unsuitable for life to EXIST. However, conditions were right for life to FORM.

In a distant place in the universe, scientists have recently also discovered molecular material for the creation of life in the so-called "Lagoon Nebula," within the constellation of Sagittarius. A molecular interstellar cloud contains all chemical components for life. One logical theory is that comets, which are originating from interstellar dust of this type, could have carried the life-forming molecules to our planet, by crashing into a very young Earth about four billion years ago.

Such discoveries do prove an important point: given the basic necessary conditions, primitive life CAN and probably WILL be created ANYWHERE in the universe. Looking at our planet's proven evolutionary development, there is no reason NOT to accept evolution of higher life forms on other planets. Given enough time and for conditions to change slow enough for life to adapt, life-forms will not only survive but some of them will eventually develop into higher, more sophisticated creatures. Once again, nothing seems to be unique to Earth, and from what we already know from other stars, nothing is unique to our star, the Sun, either.

MORE ABOUT SATURN

A Saturn year equals almost 30 Earth years, partly as a result of the much greater distance from the gravitational forces of the Sun. Therefore, a hypothetical resident of Saturn could be lucky enough to see two "summers" and "winters." The most remarkable feature of Saturn are its rings, visible under most favorable conditions through good binoculars. But it is quite an experience to actually observe this unique phenomenon through a small telescope.

Three basic rings are made up of swarms of icy particles as small as tiny crystals and as big as about 30 ft. in diameter. One could call an object of this size a "micro moon." The American space probes "Pioneer II" and "Voyager" have given mankind much valuable detail about the rings, including more precise dimensions and details on the material of the discs.

Although the outermost ring has a diameter of about 175,000 miles (compared with a more familiar distance — the Moon is not much further from Earth!), the maximum thickness of the rings is only about 10 miles. Saturn is not the only planet with orbiting ring "discs," Jupiter also has a very thin and dark ring, as well as the next planet past Saturn, Uranus and possibly also Neptune. Even Saturn is not so unique after all.

As mentioned, Saturn's atmosphere is similar to Jupiter's with very high contents of hydrogen and helium gases. At the much lower temperatures on Saturn, these and other gases must be in a liquid state as floating droplets, "fog" or "slush," compared to the invisible state of these elements in their gaseous form on Earth. The temperatures at the top of the upper cloud layer of Saturn were measured to be about 38° C colder than on Jupiter, another un-inviting place for earthlings.

URANUS, TOO DISTANT TO BE SEEN WITH THE UNAIDED EYE

Our next stop is the search for the planet URANUS. However, we now need a telescope and a lot of luck to find it. It would be easier to follow up on the results of the Voyager 2 encounter with that planet in January 1986. Until then, we knew little about this very distant planet. We could observe only four moons from earth observatories, until Voyager 2 surprised us with a total of 15 moons and a Saturn-like disc.

At a distance of 1,782,000,000 miles from the sun, it takes sunlight over 2½ hours to reach the surface of Uranus. Most of the tiny bit of the sun's rays which reach the planet, are reflected back into space. The effective temperature on Uranus is minus 200° C, but it has to be said that this figure is only relative, based on the unusual orbital position of Uranus.

Uranus is tilted 98 degrees to the "plane" rotating with the poles facing the sun. Since it takes 84 Earth years for one orbit around the sun, each pole and the upper and lower portions of Uranus have a 42-year long "summer" and constant light and later a 42-year "winter" and constant darkness. However, one must realize that at Uranus' immense distance from the Sun, 42 years of "daylight" really mean light from a bright star only.

We have come to the end of what can be observed with or without simple binoculars from our observation point at the pier. At least as far as satellites, planets and asteroids are concerned. Before we go on to the next closest "solar system," we will catch up on the latest information on the most distant planets in our own solar system.

NEPTUNE, WHAT WE SEE HAPPENED OVER FOUR HOURS AGO

The next planet in the line up is NEPTUNE at the unimaginable distance of 2.8 billion miles from the sun. To be precise, Neptune is and will be until 1999, the outermost planet of the solar system. The "usual" outer planet Pluto's unconventional orbit has brought Neptune temporary closer to the sun.

It takes sunlight about four hours and twelve minutes to reach Neptune, where one "year" equals about 164 Earth-years. Indeed a very strange world with temperatures hovering around minus 220° C, when all known elements behave different than on and around any of the planets with more "moderate" temperatures. All gases are either liquid, slush or on the verge of becoming solids on the surface of Neptune. The planet has no real solid surface as such and at about 33,000 miles diameter has over four times the diameter of Earth.

Although the astronomer Galileo may have discovered Neptune a few hundred years ago, little was known about it until much later. Sunlight reaching Neptune has about 1/1000th of the intensity of solar radiation reaching Earth and detailed observations were only really possible by the Voyager II probe.

Voyager II passed Neptune's north pole by about 3100 miles in 1989 after a 12 year flight from Earth. Neptune has eight known moons, six of them were recorded by Voyager II, the largest being the most unusual moon Triton with an atmosphere of mostly nitrogen. Neptune actually emits more "heat" than it receives from the very distant Sun, similar to the other giant planets, Saturn, Jupiter and Uranus.

Probably one of very few similarities with Earth, Neptune's core is thought to be as hot as our own planet's interior. Like on Earth, active geysers eject hot gases into Neptune's atmosphere and active volcanos would add their portion of heat into the minus 220 °C layer of atmospheric soup. Possibly the temperature contrasts created by geysers and volcanoes in the bitter cold atmosphere of Neptune are a logical explanation why the planet is the "windiest" in the solar system? Every little bit of heat helps at THOSE low temperatures.

PLUTO, USUALLY THE MOST DISTANT COMPANION PLANET, PLANET "X" AND THE ANCIENT SUMERIANS

The final known and most distant planet in our solar system is PLUTO. Until 1930, when it was first observed through the latest telescope of the day at the Lowell Observatory in Flagstaff, Arizona, this planet existed through mathematical calculations only. At a distance of 3.5 billion miles from the sun, it is a small wonder that it was discovered at all. It receives only a minute quantity of sunlight from the brightest "star" in the sky. The distance from the sun translates into 5½ light-hours and it takes Pluto about 248 Earth years to orbit once around the sun, completing one "local" year. A hypothetical human life born on Pluto would last one "season" with a lot of luck.

Pluto's diameter is a mere 1500 miles, smaller than our own moon. Nevertheless, it managed to capture a small satellite, a micro moon named Charon with a diameter of about 500 miles. It orbits around Pluto at about 10,000 miles distance once every 6.4 days, the same time it takes the planet to rotate once to complete one day on Pluto.

Although Pluto is thought to be the last planet in our solar system, there is strong evidence to indicate the existence of at least one other planet beyond. The theory is based on minute orbital variations by other planets and on astronomers' calculations. It would be simply too dark to be observable through even the most powerful telescope. However, theorists are hard at work looking for the "Planet X," by checking for faint radiation, other than light.

The flourishing Sumerian civilization of ancient South Mesopotamia existed between 3500 and 2200 B.C. and also refers to a "twelfth" planet. Before they merged with Babylonia over 4000 years ago, the people of "Sumer" had their own language and had survived over 1300 years on their own, longer than any nation of modern days. They survived apparently without modern-day religious feuds.

However, the most interesting Sumerian legend is based on their knowledge about eleven or twelve planets in the solar system. If we include the Moon, the asteroid "belt" and "Planet X" today, we do come up with twelve "planets." What did the Sumerians know 5000 years ago that we didn't even suspect until recent years, and then only with the help of modern technology?

Who taught them facts about invisible objects in the

solar system, since the asteroids, Uranus, Neptune and Pluto were all not visible to them with the naked eye? Remarkable also is that they are ONE of very few civilizations, which managed to survive surrounded by hostility for so many years and WITHOUT "modern" religions. Unfortunately again, scriptures tell us little or nothing about astronomical facts and the number of planets. And about Planet X.

INTERSTELLAR SPACE BEYOND PLUTO

At this point, we must try again to comprehend the vast distances in space, just considering it takes 5½ hours for a beam of light to travel from the sun to the planet Pluto, travelling at 186,000 miles per second. It is easy to lose comprehension when it comes to the unimaginable dimensions of the universe, and I want to give one more interesting example to make it more understandable: The Voyager II spacecraft had been launched in 1977 and had travelled about 13 years to make a close fly-by of Neptune only.

It will take the spacecraft until the year 2012 to leave the sun's magnetic field and its gravitational forces (the "heliopause"). Voyager II's speed will vary greatly and is mainly influenced by close vicinity to planets or other objects throughout the solar system. For all practical purposes, I will use 35,000 miles per hour, by itself an incredible speed for the human mind to comprehend.

About 18,000 years later, long after our children's children are gone, the Voyager will be "near" our closest interstellar neighbor, the star Proxima Centauri. However, it will be passing only within about 3.2 light-years of that star. Voyager II will be one light-year away from the sun at that time. "Only" about 300 years later, the craft will pass within 3.21 light-years of the star Alpha Centauri. At that time, it will be 1.02 light-years away from the sun and we thought, 5½ light HOURS to Pluto was a great distance!

Provided of course Voyager II will not collide with another object in interstellar space (possible, but most improbable), it will be 47.38 light-years from the Sun in the year 957,963 or close to one million years from now! It will pass the star DM+27 1311 within 6.62 light-years at that time.

Is it not fascinating, how accurately scientists can predict certain events that far ahead? In comparison, the somewhat more "old fashioned" Pioneer spacecraft will take 80,000 years to reach the star system Alpha Centauri, at a maximum speed of 35,000 miles per hour.

Astronomers have recorded 240 stars within our immediate neighborhood up to thirty-three light-years from Earth. Fifty-eight of them are similar to our sun. If the construction of a 1.5 meter diameter specially advanced design of telescope materializes, we would be able to detect planets of the size of Jupiter and right down to the dimensions of Earth in the immediate interstellar neighborhood. The imperfections in the mirror of this telescope would have to be fifteen times lesser than that of the Hubble telescope, to be able to detect Earth-like planets at such unimaginable distances. Yet, a range of thirty-three light-years in the universe is like standing next to another person in a crowded elevator on Earth.

OUR HOME STAR, THE SUN

We can now go back to our observation point at the pier. First in the daytime, with a light cloud cover over the Sun, but not enough to obstruct the view completely. Remembering the size of the full moon in a clear night, it should occur to us that both objects appear about to be of the same size. This phenomena becomes more apparent during a total solar eclipse. It is not an illusion, since the sun has a diameter of about 400 times that of the moon, but it is also 400 times as distant. Our ancient forefathers could not possibly understand this scientific phenomenon and they were probably under the miscomprehension that both heavenly object were of almost identical size.

Since the sun is the closest star for us to be observed and analyzed, we already have the most accurate scientific information about it. Remarkably, what mankind has discovered so far about other stars of approximately the same size — the main components, the temperatures and other features, all found to be identical to our sun, depending on age and size. As mentioned earlier, the Sun is not at all unique, if anything, it is a rather unimportant, small and middle-aged (main sequence) star.

A rather simple scientific device, known as the "spectroscope," breaks down the captured light from the sun into the spectral lines of elements, which are present in the solar atmosphere. By comparing the results with those from other stars, this instrument provides unquestionable proof that ALL stars are more or less of the same structure and are all going through the same evolutionary process. The proportions of various elements thus measured also provides scientists with the very accurate age of the star under observation.

As a "main sequence" star, our sun will emit energy at approximately the same level and remain relatively unchanged in appearance for another five billion years, give or take a few million years. At least, that is how long our "star" will have enough hydrogen "fuel" to "burn" (convert into helium), from its core, the thermonuclear reactor "oven." The sun is now at about the middle of its life span, after ruling the planetary system already for the past five billion years.

A BASIC INTRODUCTION TO "FUSION" OR "THERMONUCLEAR" REACTION

At this point, it is necessary to introduce to the reader some basics of thermonuclear (fusion) reaction, since it is the single most significant process taking place at all times and throughout the entire universe. Without it, there would be no cosmos and no creation, no life cycle of stars and planets.

To introduce the reader to it in very simple terms: The density or gravitational force at the core of a stellar body, such as our sun causes tremendous pressure at the centre of a star on the predominant atomic matter present, hydrogen gas. The immense pressure in turn generates unimaginable heat at temperatures of several million degrees, thereby triggering a nuclear chain reaction, by first converting hydrogen into helium and later into other, heavier elements. This conversion process represents the actual FUSION reaction, which releases massive amounts of energy in form of heat and radiation.

Fusion reaction (nuclear fusion) is not to be confused with "nuclear FISSION," which is created as energy source in nuclear power plants and by exploding atomic weapons. Nuclear fission explosions are used as a trigger for hydrogen "fusion" bombs, as the only available source of enough heat to trigger nuclear reaction for "H-bombs."

Since Einstein introduced his famous equation many years ago, $E = mc^2$ (Energy = mass, converted [c= speed of light]), astronomers had many successful opportunities to prove this and other theories. To understand the equation somewhat better, I will try to "translate" it into layman's terms: The idea is that the ordinary "resting" mass of a body is also associated with an amount of energy equal to its mass, multiplied by the square of the velocity of light. Or to make it very simple: Matter, energy and time are all interdependent with each other. Just another indication to me, and to many logic thinkers, Einstein WAS in fact a true "prophet." At least, his "prophecies" have so far nearly all been proven to be correct.

Depending on mass and density at the core of a star, fusion reaction (or thermonuclear reaction) can and will go on for billions of years, and as long as there is enough hydrogen "fuel" available for conversion into helium and other elements thereafter. I will get back to this subject, when we examine the birth, life and death of stars. At this point, it was to be a simple introduction only to basic thermo-nuclear physics.

MORE ABOUT OUR SUN

The sun's approximate distance from Earth of nearly 93,000,000 miles is often used as an astronomical distance measurement, equalling one Astronomical Unit (A.U. — not to be confused with the same abbreviation for "Angstrom Unit"). With a diameter of 865,000 miles, the Sun's diameter is 109 times that of Earth, or a one inch marble compared to a 109 in. diameter balloon. The solar surface temperature has been recorded at about 10,260° F (5,700° C) and although THAT is hot, what about the intense heat of the solar core, where the thermonuclear reactor furnace is operating at 36,000,000° F (about 20,000° C)? By the way, if one could walk on the surface of the Sun, this person would weigh nearly 28 times the terrestrial weight.

To illustrate the powerful effect of the Sun on our planet, and all other planets in the solar system, one only has to pay attention to the mysterious "sunspots," also known as "solar flares." They are best described as eruptions from huge "craters" within the gaseous solar atmosphere and with a twenty-two year high and low cycle.

Solar flares are by far the most massive gaseous eruptions in the entire solar system, having a major influence on the terrestrial weather and radio transmissions, interrupting the latter at times or at least seriously impeding them. Yet, the sunspots cover a relatively small area, compared to the total surface of the sun, even though most of them are big enough to swallow our planet several times over.

NOTHING IN INTERSTELLAR SPACE IS "STATIONARY"

Here is one important fact about astronomy to keep in mind: What to an earth-bound observer appears stationary in the sky, or at best moving very slowly, is really travelling incredibly fast. Of course, the most noticeable movement in space for us is the rotation of Earth once every 24 hours, giving us day and night. The next best known motion is the annual orbit around the Sun, although not as clearly visible to terrestrial observers. It gives us the four seasons, while spaceship Earth travels around the Sun at the incredible (average) speed of 18.46 miles per second.

But there is one other motion, more incredible, but not noticeable at all to us: the orbital movement of the Sun around the centre of our galaxy of stars, the Milky Way, at about 138 miles per second. At that speed, it still takes our Sun and all orbiting planets, moons and asteroids a full 230,000 years to complete one orbit, together with all neighboring star systems completing one "galactic" year. And finally, one other motion has to be considered: that of the entire galaxy moving through space — but I will get back to it later. The important point to remember is: *NOTHING is stationary in the Universe, it never has been and it never will be.*

OUR "NEIGHBORS" BEYOND THE SOLAR SYSTEM AND THE CONSTELLATIONS

We can go back one more time to the night sky at the pier. The brightest "appearing" star is Sirius A, with an actual brightness of about 26 times that of our sun. The readers will remember my earlier reference to this star and to it's nearly invisible twin, Sirius B: The African Dogon tribe knew about this "binary" star system long before modern, "civilized" men tried to teach other native tribes (or force upon them), their Judea-Christian-Islamic views about the universe and creation.

Without providing a detailed listing of the brightness of stars, the so-called "scale of apparent magnitude," I have instead selected some of the MOST OBVIOUSLY visible stars from the night sky. All of them can be seen with the unaided eye, provided they are above the horizon at some time of the year.

One important point to remember is what we see appearing to be two dimensional in the night sky, with all stars at about the same distance, IS NOT AT ALL OF EQUAL DISTANCE. When our earliest ancestors gazed into the heavens, they must have had the same impression as we do today, the night sky looks as if thousands of tiny lights were painted on the inside of a giant dome . . . In reality, when we observe two stars side by side, one of them may be much brighter than the other, yet the brighter one of the two may be ten times as far away as the star with the lesser magnitude. The opposite applies also — a somewhat brighter star may be in reality 100 times as bright as its "apparently" only slightly "weaker" celestial "neighbor," yet they may be separated by thousands of light-years. And they probably are.

Sometimes, a bright star really is much closer and considerably smaller than a distant apparent "neighbor," only a star map or celestial atlas would clarify the situation. As a rule of thumb, no two adjacent "appearing" celestial objects are ever "near" to each other. The exception are our planetary neighbors, while distances between stars are nearly always measured in light-years.

What is known as "constellations," and what is used for identification in astronomy as well as for astrological charts, are groups of stars, which only APPEAR to us as a close family group. Constellations are used as REFERENCE points *only* for star maps. All of them will change greatly over a period as "short" as one million years that the present picture will be entirely different by then.

If we were able to observe any of the imaginary, man-made constellations from another point in the Universe, this particular group of stars would not at all appear as a constellation. In addition, what is *seen* as LIGHT emission from a group of stars may appear entirely different when other radiation sources from the same point are recorded. Invisible sources may be radio frequencies, ultra violet and infrared light, gamma rays and X-rays, etc. The difference in strength of emission compared to visible light from the same object can be considerable, often giving comparative charts an entirely different look from what can be visually observed.

ASTROLOGY OR ASTRONOMY?

Since I referred to ASTROLOGY earlier, I want to clarify the difference between the "science" of ASTRONOMY and the "art" of ASTROLOGY. The dictionary describes "Astrology" as "the art of predicting or determining the influence of the planets and stars on human affairs." Although it is POSSIBLE that astrological forces may or may not influence our lives, as of now, we have been unable to attach any scientific proof to Astrology. It is possible that an accumulation of various radiating forces originating from any given "constellation" may cause one's life to be influenced as a "Gemini" or a "Leo" or whatever.

Astrology is a very ancient art and may have been a very important part in the lives of our Greek and Roman ancestors, particularly before "modern" religions were introduced. But like religions, astrology is NOT a science and nothing has to be proven beyond a shadow of a doubt, unlike scientific theories and findings. Until the 17th century, astronomy and astrology were closely related, due to lack of knowledge about the real universe at that time. The invention of the telescope changed all that and mankind had to come to the conclusion that amongst other startling facts, the Earth is NOT the center of the universe.

HOW IMMENSE ARE GALAXIES REALLY?

Let us go back to the pier one more time, since it is a very clear night. With the exception of two or three visible neighboring planets and one or two very faint patches of light, ALL the stars visible to the unaided eye are within our galaxy, the "MILKY WAY GALAXY." Even a pair of binoculars or a small telescope would not make much difference, the observer may be able to see one or two more tiny faint "hazy patches," which are not stars, they are actually neighboring galaxies.

Until this century astronomers thought these light patches to be "nebulae," or gas clouds and part of what they believed to be the "entire" universe, the Milky Way. It took humanity until only decades ago to accept the existence of countless other galaxies beyond our own. "Extra-galactic" distant objects are not to be confused with gaseous nebulae and "stellar clouds" within our own galaxy. The numbers and density of these cloud-like objects becomes much greater towards the centre of any galaxy, including our own.

The core of the Milky Way galaxy is obscured by this type of cloud accumulations, covering tens of thousands of light-years in area. The same applies to all other "spiral" galaxies. However, the observer at the pier does not have to be too concerned with these objects, since they are hardly visible to the unaided eye or with simple optical aids. The total number of galaxies is at latest count estimated to be in excess of 40 billion. Once again, it was science which opened this immense horizon to mankind, not religion.

Out of the twenty-five brightest and nearest stars, we can observe Alpha Centauri A, the nearest visible interstellar object at a distance of a "mere" 4.2 light-years and the giant star "Cygni" with a little bit of luck, at a distance of about 1600 light-years. Most other visible stars in the "immediate" neighborhood are located at the following distances (in light-years) and in random directions from a terrestrial observation point:

Sirius A	8.7	Canopus	98.0
Arcturus	36.0	Vega	26.5
Capella	45.0	Rigel	900.0
Procyon	11.4	Betelgeuse	520.0
Archemar	118.0	Hadar	490.0
Altair	16.5	Aldebaran	68.0
Spica	220.0	Antares	520.0
Formalhault	22.6	Pollus	35.0
Deneb (Cygni)	1600.0	Beta Crucis	490.0
Regulus	84.0	Acrux	370.0
Adhara	680.0	Shaula	310.0
Bellatrix	470.0	Elnath	300.0

The list from top to bottom is in order of brightness, better known as "apparent magnitude," with the brightest stars at the top. The obvious great variations in distances are due to the equally great variations in size and "real" brightness (absolute magnitude) of a star. One more reminder that the brightest star, Sirius A, only APPEARS that way to us and it has little to do with actual distance. The absolute nearest star, PROXIMA CENTAURI at slightly less distance from Earth than Alpha Centauri A, but it is NOT visible with the unaided eye. Only an eight inch telescope will allow the terrestrial observer to see the closest neighbor.

Proxima Centauri as the nearest celestial neighbor is of particular interest to us and some of the statistics are worth noting: At 40,000 miles diameter (five percent that of our sun), it could be compared more to the planet Jupiter, a bit too large for a planet and equally too small as a star. To some scientists, anything emitting more light and heat than it receives is not a planet. To some, Jupiter falls into the same category as Proxima Centauri, as mentioned earlier.

If we were able to replace the Sun with this miniature star, by placing it in the same spot and distance from

Earth, we would have the light of about forty-five full moons or 13,000 times less luminosity than that from our Sun. No danger of sunburn, but it would be unbearably cold on this planet.

Due to its small mass, Proxima is considered a "red dwarf." There is not enough matter and gravity available to trigger the same massive nuclear reaction as our sun. The amount of light and heat emission from Proxima is therefore very moderate.

While at the subject of dimensions of "neighboring" stars, let us compare some of them. Outstanding by far is the visible giant "Rigel" in the constellation Orion, with a true brightness of 150,000 times that of our Sun. If it was as close as Sirius A to our solar system, it would shine 100 times as bright in our sky and we could see it during any clear day.

The absolute "emperor" of the stars is too far to be seen with the unaided eye. At a distance of 11,000 light-years, the monstrous HD93129A is about 1,000,000 times as bright as our Sun.

Alpha Centauri A, Alpha Centauri B and Proxima Centauri orbit together as a multiple star system, all at immense distances. The average distance between Proxima and the A and B pair is about one trillion miles or one sixth of a light-year. It takes Proxima about a half million years to orbit once around it's bigger brothers.

OUR OWN "AVERAGE" GALAXY

The reference to our own galaxy as "Milky [Way]" comes from ancient Greek, yet it is still used today by twentieth century astronomers. The bushmen of the Kalahari desert in Africa call it the "Back bone of the Night," since it appears as a "spine" in the night sky.

We are now at the end of our observations from the pier, only extremely powerful telescopes will reveal the secrets of the depth of our own galaxy and others. The orbiting Hubble telescope has been repaired and will open up vast opportunities for new discoveries. As our atmosphere protects all life-forms from ultraviolet radiation, it also obscures the light rays from most stars and neighboring galaxies.

Astronomers are often more successful by searching the night sky with RADIO telescopes for emissions of radio waves from interstellar objects. Radio telescopes have a much greater range compared to optical equipment, since our atmosphere passes the frequencies of radio waves somewhat more readily .

It is estimated that our galaxy consists of a large "family" of stars, currently said by some to have as many as 400 billion members. Earthbound astronomers at best can only identify less than 100,000 individual stars, give and take 10,000 or so, more through long-exposure photographic recordings. It was estimated only a few years ago that our galaxy would consist of "only" 100 billion stars.

Since the solar system is located near the outer edge of the galaxy, optical and radio telescopes can only record neighboring star systems around us and reasonably close to the center of the Milky Way, but nothing really near and beyond the center. Just as most other galaxies, ours is disc or saucer-shaped and the core's density is of immense proportions.

The diameter of our galaxy is rather average at roughly 100,000 light-years across. In other words, if we could penetrate the massive core with optical or radio telescopes, we would "see" the opposite edge as it was nearly 70,000 years ago. The solar system is located one third from the Milky way's outer edge, therefore a distance of "70,000 light-years" was given as approximate distance from the "opposite" edge.

As we approach the center of the galaxy during a hypothetical spaceflight at the speed of light, space becomes more "dense." The gravitational pull from the core material captures more star systems than the outer edges. One can simply compare our galaxy to a whirlpool in the water, where the center "funnel" or "vortex" develops the greatest pressure and the most rapid motion and where it sucks large objects into a body of water. A tornado and a hurricane act in a very similar manner.

Some astronomers believe (until proven one way or the other) that the center of most galaxies is really a "Black Hole." Matter is "sucked" in through the vortex effect at such high speeds and density approaching the speed of light that all matter becomes invisible to us.

However, this theory fails to explain what happens to matter, *after* it has gone through the "funneling" effect of the black hole; does some of it accumulate near the inner edge on the other side of the funnel or does some of the highly accelerated matter continue to travel on into intergalactic space — only to be attracted by gravitation from another galaxy or a cluster of galaxies? Or does a major portion of the high density mass turn into energy during the high speed travel through the center of the galaxy? Or all of the above, a bit of everything? Maybe the Hubble telescope will provide some of the answers for earthbound scientists. I will get back to "Black holes" in a later chapter.

Reading about "density" of matter in space, one tends to compare this "matter" with "dust" on Earth, or at least with measurable movement of gases, which on Earth can be felt as wind. Interstellar gas accumulations at the highest density are still less dense than the most perfectly man-made vacuum on Earth. Once again, we must remember that interstellar distances cannot be compared with terrestrial or planetary dimensions.

It is remarkable however, how similar matter behaves in a whirlpool or in a hurricane on Earth, or in the vastness of outer space. A tiny whirlpool created by a paddle in the water appears to react in the same manner as a hurricane, photographed from a satellite. A distant spiral galaxy behaves in the same whirlpool pattern and motion, and examples of spiral galaxies can readily be observed through powerful telescopes.

Illustrated: Hurricane Andrew, partly photographed from the GOES 7 satellite and partly simulated. Also, a photograph of the Messier 74 galaxy of stars, (122.5 million light-years distant). An example of the similarity of forces in nature, regardless if a terrestrial "spiral" formation has a diameter of a few hundred miles or if a galaxy has a diameter of 100,000 light-years. The "Funnel Effect" — is it coincidence?

Let us look again at the unimaginable number of stars in our own Milky Way galaxy, 400,000,000,000 or so at latest "count," to be about as precise as we can get. All stars appear to go through the same predictable life cycle, the thermonuclear reaction is the same, and is determined only by their ages and by their often vastly different dimensions. Over 50% of stars within our galaxy orbit with a companion as a "binary" system or as a multiple star system. Our sun is in the minority, unless one considers the much smaller "planet" Jupiter as a companion of the Sun.

Some scientists have reason to believe that the mass of all planets in the solar system originally was combined in a "twin" star of the Sun and later broke up through some unknown, but obviously quite catastrophic event. If that has happened, it would have taken place in the very early stages of the solar system's creation, billions of years ago. Astronomers will probably never be able to prove or disprove this theory. In any case, this imaginary solar twin would have been very tiny, since the Sun holds 745 times as much mass as all planets of the solar system combined.

Only very recently have astronomers been able to calculate the existence of large invisible objects (planets?) orbiting other stars by measuring minute deviations in the motion of the "parent." It may be the first proof of the existence of planets outside the solar system. A large planet, such as our own Jupiter, would have this effect on the orbit of any star. Therefore, indications are again that our solar system is in no way unique.

It is becoming ever more probable that most other stars indeed have a captive planetary system of some kind. The systems may be as diverse as our own, where dimensions vary from the tiny planet Mercury to the giants Jupiter and Saturn. We have no reason to believe any longer that the solar planetary system is one of a kind. Scientists have long discovered how "average" our sun is in every respect compared to billions of other "sister" stars. That is billions, with a "B."

With up to 400 billion stars forming the Milky Way galaxy, numerous estimates have been made about the number of planets with similar characteristics as Earth, orbiting in other "solar systems." The most optimistic estimate is one per star, since our solar system has Earth as the one planet, which is reasonably comfortable for sustaining a multitude of life-forms. With this optimistic calculation an overwhelming 400 billion earth-like planets would be part of our galaxy.

Along with most astronomers, I tend to be more sceptical by giving a much more conservative number a realistic acceptance. Let us say, only one out of one hundred stars has an earth-like planet, it would still give us the unimaginable total of four billion sister planets. Let's now be really pessimistic by assuming that only one out of one hundred of THESE planets is of a similar age as Earth, with evolution at about the same stage of development. We still end up with 40,000,000 sister planets in the Milky Way Galaxy only.

In addition to visible and invisible stars on the "other side" of the galaxy, astronomers have only now discovered evidence of so-called "MACHO" stars, just outside the outer perimeter of the galaxy. So far, only four such objects have been recorded in the southern hemisphere, after their presence had been suspected for the past twenty years. One reason for suspecting invisible matter outside of the galaxy's outer edge was the relatively similar orbital speed of stars near the edge, compared to those near the centre of the galaxy. If the "edge" really was the "end," the "outer" stars would travel much faster than the "inner" ones.

Macho stars are expired stars, with either very weak light emission or none at all, but of immense density and mass. The machos registered are between three to thirty percent of solar mass and could be considered the remains of small stars or large planets. Machos are not to be confused with the visible and larger white dwarfs. A great abundance of machos is suspected outside the outer limits of the galaxy.

To come to the point: our galaxy may possibly be made up of a much larger number of stars (and planets) than the currently estimated 400 billion.

The most pessimistic astronomers estimate the number of planets similar to our own within the Milky Way galaxy to be as few as 50,000. It would really mean, they expect only 0.01% of the stars in the galaxy to be capable of supporting a planet like our own. Still a remarkable number of planets, where humanoid creatures could be hundreds of thousands of years AHEAD of us in technology or where they could be the same number of years BEHIND earthlings. This subject came up earlier also.

As mentioned, a period of 100,000 or even 200,000 years of a planet's life of about 10 billion years compares to a few seconds in a human life. But a lot has happened to the higher life-forms on Earth over a "short" evolutionary period of 200,000 years, as everybody has to admit.

Taking this most conservative estimate into account, it is almost impossible NOT to accept advanced humanoid creatures on all 50,000 or at least on most of these planet. The circumstantial evidence is simply too overwhelming, not to accept this conclusion as simple fact. I can only attribute a sceptic's objections once again to religious teachings. To think of the possibility of at least 50,000 or four billion inhabited planets within our galaxy alone is simply beyond the traditional believer's comprehension, and not at all in compliance with religious teachings.

Let us recall again: WHEN the scripture was written, at a time when mankind truly believed that the (flat) Earth was the center of the universe and a few thousand visible tiny dots, called "stars," were all there was besides the Sun and the Moon. Stars were pretty little things of unknown dimensions and distance, glimmering in the heavens, something to wish upon.

Working on the scrolls for the Bible 3000 years ago, how could the authors have known about the existence of 400 billion stars in our galaxy alone? And with the probability of between 50,000 and four billion earth-like planets orbiting around them? Most ancient civilizations had no knowledge that Earth was orbiting around the sun, without any idea about the controlling forces of microorganisms upon all life forms. There was no knowledge about the nearly identical behavior of all matter in the universe. Much of this information was only acquired by humanity through science over the past century, with a lot more to be discovered.

With the nearest star system at a distance of up to four light-years from Earth, the obstacles to interstellar space travel by terrestrial astronauts seems at times impossible to overcome. At conventional speeds of today's spacecraft, it would be indeed more than an obstacle, it is simply impossible. As noted earlier, unmanned space probes such as Pioneer and Voyager travel 18,000 to 35,000 years just to get "close" to the nearest star system.

How then are alien visitors travelling to visit our solar system? Having referred to it briefly earlier and according to one of Einstein's theories (proven by now), for any object inside or attached to a spacecraft travelling close to the speed of light, *time slows down* accordingly. While well over four years would pass on Earth, the astronauts in a spacecraft capable of accelerating to about 180,000 miles per second, would only take a few months to arrive at our nearest interstellar neighbor, according to THEIR clocks.

They would only age a few months, and a few more months upon returning back to Earth, while the round-trip would probably take close to ten earth years! With our current technology, we can only imagine acceleration of a spacecraft to nearly 180,000 miles per second. Today, we are doing good, if we reach 35,000 miles per hour, less than a "mere" ten miles per second. Planet Earth orbits around the Sun at an average of eighteen miles per second, in comparison. In theory with yet to be developed fuels and/or thermonuclear powered engines, man-made space craft CAN and will ultimately reach the speeds required for interstellar space travel.

Illustrated: The "Andromeda Galaxy," one of our "nearest" neighbor galaxies and one of at least forty BILLION. Its diameter is about average. To cover the distance from one visible edge to the other, light travels about 100,000 (light) years at 186,000 miles per second. Individual stars are not visible from our solar system (over two million light-years distant). What appears as single stars, are either deep layers of stars, interstellar dust, small companion galaxies or stars from or own galaxy, all of it separated by many light-years.

The Sirius multiple star system at 8.7 light-years distance could also be reached in a matter of a few more months. Unfortunately, most of the scientists and technicians involved in the original launch of the spacecraft would either be dead or about thirty years older, when the space travellers would ultimately return from their return journey to the Sirius pair, unless they could survive in suspended animation. One way or the other, it is therefore quite probable that the Dogon tribe in Africa indeed had visitors from that part of our galaxy, many years ago.

GALACTIC NEIGHBORS, AND THEIR ORIGIN

During a final look at the night sky from the observation point at the pier, we may be lucky enough to see one of those hazy patches of faint lights, which I referred to earlier. Other than a few interstellar dust "clouds," nebulae and star clusters within the local galaxy, only one neighboring galaxy is visible outside our own. It is the only object distinguishable to the naked eye that is not from inside the Milky way.

I am referring to our closest major neighbor galaxy "M31," or better known as "Andromeda." The visible and invisible radiation from this galaxy reaching us today left 2,000,000 years ago. If there were any major changes to Andromeda during the past few years, we would not find out until two million years from now. And that is a CLOSE neighbor galaxy.

Andromeda is a "typical" spiral galaxy, a bit smaller than our own and part of a local "cluster" of twenty galaxies. Nothing seems to be out of the ordinary, when astronomers observe Andromeda. It is proving once again, the universe is made up of the same material and stars behave basically in the same way everywhere. There are of course a number of exceptions to this rule. One comes to mind immediately and it needs some explaining: Astronomers have noticed some time ago, that all interstellar objects are moving away from each other. In other words, no matter where we are located in the universe, our neighbors are moving away from us.

To illustrate this phenomenon, one has to imagine hanging suspended inside a huge balloon with little helium-filled balloons all around the subject, all of them suspended on strings. Someone is inflating the large balloon and the inside space of it expands. At the same time, all the little balloons above and below, left and right and in front and behind move away from the suspended observer.

The small balloons are inflated also and contain small suspended particles. They represent the expanding galaxies and the particles inside are the individual stars, moving away from each other as the small balloons are further inflated. The point is: All galaxies and stars, with a few exceptions, are constantly moving away from each other everywhere in the observable universe.

Andromeda is one such exception, it actually moves TOWARDS our galaxy, at about twenty-two miles per second. Nothing to worry about — even if Andromeda would move at the speed of light, it would take two million years for a collision to start with our galaxy. Considering the vast distances in space, it would be highly improbably for any of Andromeda's stars to collide with some of the Milky Way's interstellar objects. Possible, but highly improbable. Collisions between distant galaxies can be observed now and are not unusual but do not appear to involve frequent collisions between individual stars.

Generally, colliding galaxies are re-shaping each others structure, but collisions between stars are really very rare. With Andromeda approaching at twenty-two miles per second, the actual process of "collision" will take somewhere around 100 million years to complete, for all stars from both galaxies to integrate into a newly formed, larger galactic system.

Still referring to Andromeda, not long ago it was thought

to consist of 100 billion stars, but now estimates are much higher. With a diameter roughly the same as Milky way (100,000 light-years across) and since more powerful optical telescopes and radio telescopes are now available to astronomers, one would logically expect somewhere close to 400 billion stars in the neighbor galaxy also. Both galaxies are slightly larger than average, with most of the others in the local group being much smaller. That is, if one can consider any galaxy as being "small" — Andromeda also has two truly "small" companion galaxies.

The difficulty in estimating the number of stars making up Andromeda is the fact that individual stars cannot be recognized. What appears as a hazy patch are really millions of individual stars separated by light-years. All of it is still hard for the human mind to comprehend.

From now on numbers may become meaningless to the reader. When it is hard to imagine a distance of two million light-years to the nearest galactic neighbor, how can one comprehend distances of fifteen billion light-years? It is the time it takes light to reach us from the most distant observable or measurable galaxies.

The distance of "only" 100 million light-years is really hard to grasp — it is the "diameter" of our local "super cluster" of galaxies. Our LOCAL cluster of twenty galaxies is part of a super cluster and is located at the edge of it. The core cluster "Virgo" is by far the largest, located about 50 million light-years from ours. This cluster in the direction of the constellation Virgo alone consists of THOUSANDS of member galaxies.

The predominantly expanding motion of the universe has been noticed by astronomers through a "red shift" of visible light. A red shift can be compared with the changing tone of the whistle from a moving train. As the train approaches, the whistle's tone appears at a higher pitch and after it has passed by, the sound appears to be of lower frequency to our ears. It "appears" as a shift to a lower frequency. Measurable light from distant galaxies also appears more reddish, being the lowest of frequencies in the light spectrum.

The theory of an "exploding" universe, the "BIG BANG THEORY," originated from this known expansion (red shift). Most astronomers have settled for the time being with the Big Bang theory, others go a little further, by believing in a "pulsating" universe. I tend to lean towards THAT theory. Maybe it seems to be more logical.

By pulsating, believers in this theory also start with the Big Bang theory, but expect the universe to contract again, when the gravitational forces between neighbors become stronger than the outward forces from the last "Big Bang." As the billions of galaxies in the universe get closer and closer to each other again, intergalactic space debris, gases and all other matter is pulled more and more towards various super clusters, which in turn are pulled closer and closer to each other. In other words, the expanding motion has reversed itself and mass is accelerating towards a central point (or several of them) in the universe.

As the density increases, the number of collisions will increase, and the higher density generates increasing temperatures to the accumulating matter, into billions of degrees and in billions of thermonuclear furnaces. Ultimately, the entire matter of our universe will be packed into one gigantic super-nuclear furnace. With an evermore rapidly increasing density at the core, another super thermonuclear explosion will cause the next "Big Bang" and the whole process starts all over again. At least, that is the most logical theory to me, to many other amateur astronomers and many professional scientists.

The pulsating process could repeat itself several times, until most of the vast amounts of hydrogen have been transformed into heavier elements. After that, who knows what would happen to a universe of helium and mostly heavier elements. The possibilities almost get just as hypothetical as the Big Bang theory itself.

Unfortunately, we will not be around long enough to follows this theoretical event, and if we could, all life forms would long be gone and re-converted into superheated basic elements. Once again, the reader will not find a prophecy about all of it in the Bible, the Koran or any other ancient holy books..

Dimensions become simply mind-boggling, considering that there is no known centre or "end" of the universe. Astronomers can observe galaxies in any direction from here, over 10 billion light-years distant and moving away from us at up to about 90% of the speed of light. Up to now we had established an artificial limit of up to twenty billion light-years for the "measurable" universe, where the (red shift) expansion reaches the speed of light as seen from here. With the latest information obtained from the Hubble telescope, the age of the universe is now estimated at less than 10 billion years. Therefore, the previous theoretical limit of the universe at 20 billion years must be off. But then, when we think in terms of "billions" of years, it really makes little difference if we are out by a few billion.

Getting back to reality, no matter in which direction astronomers search, everything appears basically to be same. Except with increasing distance, the more distant parts of the universe are much younger. What we "see" happened as far back as ten billion years ago, and astronomers have discovered a number of remarkable phenomena, unknown in the immediate neighborhood of the "old" or "present" universe. One such phenomena are "Quasars," best identified as extremely distant "star breeding grounds." More about them in another chapter.

CONTINUOUS CREATION IN THE UNIVERSE, OBSERVABLE AND PREDICTABLE?

Modern day astronomers brought us to a live theater, showing us the birth, the life and the death of stars in the ever-changing cosmos around us. Most of us may not realize it, but we are the most fortunate generation in the history of humanity to be able to witness actual continuous creation in the universe. We are part of it and we live in the middle of it. Yet our "star" and the entire solar system, are only an insignificant, but typical tiny part of an enormous system of unimaginable dimensions, the UNIVERSE. We no longer have to rely on mythical scriptures, written by ancestors, who through no fault of their own could not possibly have known what REALLY happens out there in the great depth of space.

Astronomers can observe stars in a variety of stages of development along a precise evolutionary path. Much has to be discovered yet and the results have to be interpreted. The outcome is often controversial creating a variety of new questions. However, modern astronomy and astrophysics have only had a chance to seriously explore the unknown mysteries of the cosmos over the past thirty to forty years with the help of computers and more powerful optical telescopes, man-made satellites and radio telescopes. The most recent discoveries have been so overwhelming and the job of analyzing will have to carry on for years, just to process all of the obtained data.

The past forty years have been a mere split second in the history of humanity compared to the most recent 10,000 years of recorded human history. Attached by gravity, we are "stuck" to the surface of an insignificant speck in the cosmos and protected by a comparatively paper-thin layer of atmosphere. It allows surface-bound creatures to live on this planet in relative comfort, on the comparatively very small inhabitable land areas.

We are conditioned by our surroundings to the point that we cannot comprehend, what the universe really is — how we are controlled by the slightest changes out there in the vast vacuum of space, at temperatures all around us near absolute zero. Or as it is known to scientists, near "0° Kelvin" (about minus 273° C).

THE CREATION OF A STAR

Still on the subject of "Continuous Creation", let us look at the observable creation, the life cycle, the death and after-death of a star, such as our own, the Sun.

In the beginning are dark interstellar dust clouds, astronomers can measure their existence, but they cannot always "see" them. The "clouds" can be found throughout our galaxy. However, they are not at all comparable with the high density, earth-bound clouds of water vapor. Compared to even the most invisible gaseous cloud on Earth, the interstellar accumulations are nearly a vacuum. They can be observed as a nebula through reflection from nearby star clusters, or as dark clouds blocking out visibility to certain areas of the galaxy, which are normally populated by an average number of stars. Radio telescopes only can monitor certain activity within interstellar dust clouds.

Nearly without gravitational interference from other "nearby" objects in open space, the drifting atoms from (mainly) hydrogen gas clouds attract each other and start to accumulate into small clusters, through their own individual minute gravitational forces. Once the density of the "globule" becomes high enough with about 1000 to 2000 particles per cubic centimeter, the cloud will start to collapse resulting in further increase in density. As a comparison, "empty" interstellar space has a density of about one atom per cubic centimeter, while the density of globular clouds is about 300 atoms per ccm, just before the collapse starts.

In comparison again, the air we are breathing has a density of 100 million trillion atoms per ccm. Gravity is pulling air down towards the Earth's surface, contrary to matter in deep interstellar space, where only the interaction of sparsely scattered atoms is creating minute amounts of gravity. Unfortunately, it is nearly impossible for future space explorers to confirm this action, since nearly gravity-free interstellar space is too distant for exploration. In any event, the visiting spacecraft would

Illustrated: The Pleiades: Visible in the constellation "Taurus," a cluster of about 250 very "young" stars and developing planetary systems, still partially surrounded by interstellar dust "clouds" (only two to fifty million years old). Five to seven stars (so-called blue-white giants) are visible to the unaided eye at a distance of about 400 light years.

disturb the area around it in space through the craft's own gravitational pull.

Scientists are not exactly sure, WHEN the process of collapse starts and what triggers it, except it is predictable to happen at a pre-determined time period. One such collapsing cloud has the shape of a horse's head and is visible in the constellation Orion. This large formation is called "Horsehead Nebula" and it partially blocks the view of a more distant region, the HII region in the galaxy. The Horsehead Nebula stretches across a vast area of several thousand light-years in our galaxy.

During the process of collapse, and as a result of increasing density and speed of particles, temperatures are starting to rise. Various types of radiation begin to be generated as a result of growing numbers of collisions between atoms. Astronomers had to come to a complete interaction between observations and their own theories to end up with a nearly complete picture of stellar evolution, and to provide other sciences with this information. There is no longer any doubt, stars are born out of breeding grounds of interstellar "dust." Typically, "new-born" stars exist near regions in space, where dark dust clouds are dominant.

As a stellar cloud globule contracts more and more, it begins to rotate slowly. Previously, I referred to the similarity in vortex or "funnel" motion between a hurricane cloud formation, most galaxies and right down to a small man-made whirlpool. As another until now unexplained natural phenomena, "unequal masses" seem to revert to this whirlpool type of rotation, regardless of where the "masses" happen to be located.

The high and low pressure differences in the atmosphere during the formation of a hurricane create two such unequal masses. Another apparent law of physics seems to determine the speed of rotation, as the rotating cloud gets smaller and denser, the speed increases, just as the wind speed near the eye of hurricane.

Our sun rotates once every twenty-eight days, but it may have taken thousands of years in the beginning to

Illustrated: The Horsehead Nebula — dark interstellar dust clouds, stretching over hundreds of light years, covering other illuminated clouds and star systems in the background.

complete a solar "day" during the formation of it from a stellar cloud. The entire process for a star to form from a rotating, collapsing stellar dust cloud to a glowing globule and on to a PROTOSTAR, takes hundreds of thousands of years, a relatively very short time period in astronomical terms.

The time period depends strictly on the amount of mass available from the original dust cloud. The apparent brightness of a protostar can be hundreds of times as much as from the "finished" star's emission during its main sequence. As a protostar our sun may have been about 300 times as bright to a hypothetical distant observer, compared to today. Before "main sequence" (adulthood) of a star and the period of the least changes to it, it may "appear" 300 times as bright, because it is about 300 times as big in diameter, yet has a mass almost the same as the "adult" star.

As could be expected, the centre of a protostar is at first considerably hotter than the outer regions. Not enough to trigger thermonuclear reaction, and to create the stability necessary for a main sequence star. At this point, heat escapes again towards the outer region, simply because heat rises in space as it does on Earth. As a result of it, the temperature and pressure at the core area once again decreases, later causing an increase in weight (pressure) from the outer shell towards the centre.

Now the gravitational pull (read "weight") increases further, together with pressure and heat. Some of the hot gases fail to rise to the top and the weight and pressure take over. The only heat escaping at this point is through radiation and a point in the star's infancy is reached, when the so-called "radiative equilibrium" becomes a domineering factor.

After another few hundred thousand years (depending on size of the protostar), the pressure on the core and the collapse of mass towards the centre cause the inner core temperature to continue increasing to about four million degrees K. The star has almost collapsed to its final "adult" dimensions. The nuclei of hydrogen atoms are now moving more rapidly, resulting in increasing collisions and adhesions, "nuclear fusion" ("thermonuclear reaction") is taking over. Later, with core temperatures of about twenty million degrees K inside our sun, atoms are constantly converted. The heat forces the Sun to expand, yet gravity will prevent this expansion to go beyond a natural limit, during billions of years of stellar main sequence.

To reach a stable main sequence age for a star, similar in size to our Sun (one "solar mass"), an impressive thirty million years have gone by. But it is very little time compared to the star's life expectancy of eight to ten billion years (8,000 million to 10,000 million). A star of ten solar masses will reach this point in it's life after only one tenth of the time, in about 3,000,000 years. The life cycle of a very massive star is therefore much shorter than that of its smaller neighbors'. The chances for higher life-forms to evolve on its captive planets are equally too short, in this type of a "solar" system.

Astronomers can observe the birth places of several very young stars in the "Great Nebula" in the constellation Orion, with four youngsters of only about 100,000 years of age. The "Trapezium," as this "four-star-nursery" is known, lights up a large stellar dust cloud around it and has been observed since the early sixteen hundreds; only in those days nobody knew what it really was. Other star nurseries, too numerous to list, can be observed throughout our galaxy. The "Pleiades" at a distance of 400 light-years in the constellation Taurus, are also a birth place of stars. As this family of stars ages, they will separate greatly from one another. The cluster of about 100 young stars within a space of about ten light-years can easily be observed with a pair of binoculars.

During the early stage of a star's main sequence, orbiting chunks of matter right up to the size of major planets will collide with each other at random. The surfaces of the planet Mercury and our moon, are only two examples of massive, visible collisions during the "childhood" of our solar system. The surface was still hot and near-liquid, when the majority of craters formed and while there was still an abundance of objects within the gravitational forces of the moon and Mercury.

Some of the collisions are so powerful that portions of matter from colliding objects are ejected clear from the orbit of the star. Some such interstellar bodies are thought to have survived until today as comets, captured by the solar system's gravity and in orbit for a long time to come.

WHAT HAVE WE LEARNED FROM "HALLEY'S" COMET?

In March 1986, when two Vega spacecraft transmitted the image of the nucleus of Halley's comet, the information coming from a distance of about 100,000 miles was startling; it provided much more information on the basic relative equality of all interstellar matter, more than scientists could have ever hoped for. The comet's nucleus was basically composed of *muddy ice* at a diameter of up to ten miles and at a calculated weight of close to 200 billion tons. Although the comet loses up to forty-five tons per second to evaporation during a close encounter with the sun, it can afford to lose it. The comet comes close to the Sun only once every 76 years and at that rate has about 50,000 years of life left.

Some comets are said to be the only objects to visit the solar system from interstellar space. This time, scientists acquired proof of the probable existence of water and other common terrestrial elements possibly originating from other star systems. In reference to that, scientists have recently discovered "polycyclic aromatic hydrocarbons" in interplanetary dust. This important family of organic molecules may be the oldest matter of its type in the solar system and may have contributed to the formation of life on Earth, together with other essential chemicals.

Just as astronomers can predict precisely when Halley's

comet returns to the solar system, two scientists and an amateur astronomer have discovered the comet Shoemaker-Levy 9, named as usual after the discoverers. In this case Carolyn and her husband, the U.S. Geological Survey planetary scientist Eugene Shoemaker, and the amateur astronomer David Levy discovered the comet on March 25, 1993. Most significant about this comet was the violent death of it. It collided as expected with the planet Jupiter in July 1994 and scientists will have years of research ahead of them from the information obtained. This once-in-a-lifetime spectacular event was photographed from several angles and it turned out to be even more sensational than expected.

Fortunately for us, this comet did not collide with our planet, it would have caused quite a mess to the climate of Earth, to say the least. Shoemaker-Levy 9 already broke up into about twenty pieces in July 1992 when it came too close to Jupiter, causing it to orbit around that planet as a "string of pearls." The largest of these "pearls" was estimated to be about six miles in diameter (ten kilometers), about the size of object that is said to have collided with Earth about 65 million years ago. Some fragments broke up further during entry into Jupiter's top layers of matter, but most of them crashed into the large planet, causing impact markings the size of Earth in the rather strange, cloudy "atmosphere" of the planet. Whatever the damage was, with no intelligent life on Jupiter and no surface water, there can be no "sinners," who could be "punished" through a "biblical" Armageddon on Jupiter.

AGING OF A STAR

As the Sun entered into main sequence of her life, most major collisions with nearby orbiting objects had taken place already. Proof of it can be found in the much reduced numbers of impact craters on the "hardened" visible surface of planets and moons. Once the surfaces had hardened, volcanic activities were at a peak, since the thin outer crust was comparable to an ice-crust, alternately hardening or breaking up from a heavy sea.

Although there is still only circumstantial evidence of planets orbiting around neighboring stars, there is no reason NOT to conclude, that other "solar" systems have gone, or still are going through the exact same process. As noted earlier, one accepted theory is that stars of greatest diameter and mass have less chance of keeping as many planets in orbit as the sun. The reverse applies to smaller stars, since their lower gravitational forces prevented orbiting objects from colliding with the "parent" star during the time of formation of the system.

As can be expected, binary and multiple star systems are involved in a greater number of collisions during the early age of the "family." If a planetary mass does not collide with one star, chances are, it will eventually collide with the twin.

To sum it up: Once the collapse of a protostar has come to an end, nuclear fusion takes over and stellar main sequence starts. It may last only a few hundred million years for the giants, eight to ten billion years for the mid-sized stars such as our sun, and more than ten billion years for smaller stars. The number of interplanetary collisions decreases during main sequence and temperature, mass and dimensions of the star remain relatively unchanged. The conversion of hydrogen atoms into helium remains more or less at a stable rate. At the same time the energy output is also fairly consistent.

Throughout main sequence, a mid-sized star may have a core temperature of about 16,000,000° K. From there on outwards, the temperature changes sometimes drastically and in layers. While the stellar "photosphere" at a depth of about 100 miles from the surface is at a "cool" 4,500 to 5,500° K, the inner "corona" a few hundred miles above the surface may reach a sizzling hot 1,000,000° K.

At increasing distance from the surface, the temperature of the outer corona increases further to about

2,000,000° K. The drastic change of temperature causes violent stellar storms, just as drastic temperature changes in the Earth's atmosphere usually result in hurricanes and tornadoes. "Coronal streamers" have been measured to blow atomic particles into space at speeds of several thousands miles per hour.

As nuclear fusion takes place within a stellar core, hydrogen is converted to helium at precise temperature and pressure levels. Energy is continuously released and as the star ages, the core becomes hotter (also due to an increase in gravity), helium begins to convert into carbon. Fusion reaction only occurs when temperatures of about 16,000,000° K are reached. The process eventually forms other elements, such as oxygen, silicon and iron, in that order.

Much later, when the core is entirely transformed into iron, the internal heat may reach even more unimaginable temperatures, as high as five billion degrees Kelvin. The star has now reached a stage, where hydrogen is scarce and the energy release from the fusion reaction decreases dramatically. However, I am getting a bit too far ahead of myself, we are still looking at a "living," main sequence star!

One important point to remember is that throughout its life cycle every star must fight the very force, which was responsible for its creation in the first place: GRAVITY. As the gravitational forces of interstellar dust and gases pull each other together to form the developing star, it shrinks to the final main sequence dimension. Nuclear fusion, the force joining lighter nuclei to form heavier matter, is the counter-action to gravity, as it releases vast amounts of energy.

Nuclear energy, as heat, and other types of radiation, rises outward from the star's interior, thus opposing gravity. In a way, it appears to be a familiar action and reaction compared to life on Earth: powerful forces within fighting each other throughout the existence of a creature. For living organisms, the "powerful forces" that are giving, maintaining and taking life are non other than the tiniest micro organisms, atom-sized chemicals and also include life-giving (and taking) solar radiation. Yet life cannot exist without micro organisms. A star cannot exist without the opposing forces of nuclear fusion and gravity.

The most recent calculations put the proportion of hydrogen of ALL combined matter in the universe at about 92%. Helium accounts for seven percent and ALL OTHER elements account for one percent only. When a star has reached the end of main sequence with hydrogen atoms vastly outnumbered by other atomic matter, nuclear fusion is no longer a stable supply of energy. The star has shrunk considerably by that time and other elements (mostly helium) have taken over from hydrogen.

The core continues to collapse, as the heat increases from nuclear REACTION between heavier nuclei, rather than through nuclear FUSION from hydrogen into helium atoms. Even from the time of complete depletion of hydrogen at the core, the stellar "atmosphere" is rapidly expanding. The star no longer shines as bright when it expands to become a red giant, a process which takes up the first part of the last billion years of stellar life.

When our sun reaches that point in about four billion years, it will expand to a maximum of about 400 times the present size. Even before this point is reached, the climate on Earth would be unbearably hot for any living organisms to survive through adaptation and evolution. Today, Venus at up to 800° C atmospheric temperature would compare to an icebox with the solar heat which will engulf all inner planets, just before the Sun reaches maximum diameter as a red giant. During the final stages of the formation of a red giant, the diameter will have increased from the present 870,000 miles to over 62,000,000 miles. Mercury will long have been evaporated and become part of the solar atmosphere and Venus would be burning up soon after.

Earth at maximum expansion will be so close to the Sun, that the sky is bright red and our planet's surface temperatures will have reached several thousand degrees.

All water right down to the deepest oceans will have long evaporated and the planet's surface would be once more as hot as it was shortly after it was formed. The only ocean left will be that of liquid masses of terrestrial heavy elements, until then in solid form, while others have already evaporated into space. However, the entire process will not happen over night, it will take millions of years — enough time for remaining intelligent life on Earth to head for cooler planets.

Although the sun will expand to a point where the upper atmosphere almost touches Earth, the quantity of solar "mass" hardly changes. If anyone could have survived until that time, he could observe the solar core right through millions of miles of red hot gases. Solar density will be less than that of our atmosphere today. As the expansion of the red giant continuous, the density decreases further along with the composition of the gases, core density and diameter changes.

Any dying star with a mass of up to four times that of our sun will at that point convert helium into carbon nuclei at an accelerating rate. Since the relatively steady hydrogen nuclear "fusion" reaction of the core is no longer present, a so-called "triple alpha process" has taken over, causing the helium-to-carbon conversion. Eventually, this process leads to a "helium flash," near the end of the runaway fusion reaction.

Hydrogen is then only present as an outer shell and the core collapses to ever-increasing density. Various lighter elements, such as oxygen are ejected into space, and together with the remains of hydrogen, will eventually form new interstellar clouds and birth places for new stars. Escaping oxygen gas at that point will cause a greenish tint to the red giant.

After the helium flash, the dying star is left with an extremely dense core and the lighter elements continue to expand into a ring-shaped cloud, a "planetary nebula." In our solar system, the planetary nebula will continue well past the outer planets, which had been spared from being burned up by expansion of the red giant sun. The remaining very dense star is now no more than a helium burning shell with a core of mainly carbon, and a surface temperature of about 50,000° K. It will have reached "white dwarf" stage, and what is left of the sun will finally have shrunk to about half the size of Earth. By the time the white dwarf stage is reached, the sun will be very similar to Sirius B.

After another few billion years, white dwarfs eventually turn into hypothetical black dwarfs, which are totally invisible. However, astronomers have no physical evidence of black dwarfs. It has been suggested that the universe simply is not old enough yet for black dwarfs to have evolved. Recently discovered dark matter just outside our galaxy may or may not have "black dwarfs" amongst the "macho" stars. About 500 white dwarfs are known to exist within the neighboring part of the galaxy, and it would be interesting to discover planets within any of these systems.

Earlier, reference was made to the end of some stars through a violent explosion, called "Supernova" or "Nova" for smaller stars. The mass of a dying star is the determining factor; if a star has a mass somewhere between four to eight times that of our Sun, the density of the core becomes so enormous that temperatures of about 600,000,000° K may be reached. Since there is no hydrogen or helium left at that time, the remaining atomic mass of predominantly carbon starts to fuse into oxygen. This reaction causes ever higher temperatures over a very short time period, resulting in a cataclysmic reaction at the core to the point of explosion, a "supernova."

By then almost all remaining matter is thrown into space, leaving no core material to speak of. As a good demonstration of the immense energy created by a "supernova," astronomers have discovered this type of massive explosion within distant galaxies. The brightness has been of the same intensity as all other stars of the distant galaxy combined. A supernova within our own galaxy would be visible in the daytime sky, even if it was hundreds of light-years distant from Earth. Stars with more than eight solar masses create an even more powerful supernova, when their mainly iron core explodes with the brightness of a BILLION stars.

Scientists have not recorded a supernova explosion for the past few hundred years within nearby areas of our own galaxy. The remnants only of a distant supernova are clearly visible today and have been for centuries, as a gigantic bright ring "cloud." Some supernovas are visible for as long as a million years, spread out over an area of hundreds of light-years. However, our own sun it too "tiny" to even cause the "fireworks" of a nova, its will simply expand into a red giant and later shrink into a white dwarf. And in the end, all that may be left is a dormant black dwarf, smaller than Earth.

The last known and visible supernova in our own galaxy occurred on July 4th, 1054 and was than recorded in China as a "guest star." The event dominated the distant constellation of "Taurus," yet appeared almost as bright as the full moon. It was visible in daylight for as long as three months.

Astronomers were actually fortunate enough to observe a supernova in most recent years in another galaxy. Although it happened 170,000 light-years ago, it was observed on February 23rd, 1987 and was visible in the southern hemisphere only. Coming from within the Large Magellanic Cloud — a "miniature" galaxy of irregular shape and with a diameter of (only) about 40,000 light-years — it was so powerful that it could be seen by the naked eye at this distance.

A star catalogued as Sanduleak -69°202 turned into supernova SN 1987A in a matter of seconds. It started out with a mass of twenty times that of our sun, as a good candidate for a supernova. Seven years before the fireworks, the core had reached the unimaginable temperature of 1.5 billion degrees K, consuming the neon gas, which had been converted earlier from carbon. In the end, the star which had started out to be 50 times the sun's diameter and about

100,000 times the sun's luminosity, collapsed into a neutron star of only 15 miles in diameter.

Along with the visible light, x-rays and other radiation from the same source was recorded on Earth at about the same time. During the height of the expansion as visible light, the diameter was over a number of light-years, before it collapsed into a density equal to a half-inch marble containing all matter of the Sears Tower in Chicago, weighing about ten billion pounds!

The ancestors of the Hopi Indians in North America passed on the legend of the supernova, which is visible now through telescopes as the "Crab nebula," a large distorted circle of gases. However, since the supernova occurred at a distance of about 6500 light-years from Earth, it really happened approximately 7500 years ago. X-ray images taken indicated a "pulsar" core of extremely high density. A pulsar is a highly magnetic rotating neutron star, the core remnants of a star of much larger dimensions as our sun.

Surviving planets orbiting around a white dwarf today with humanoid inhabitants could be about six billion years ahead of us in technology and scientific knowledge! It will take our sun about six billion years to become a white dwarf. Survival in the solar system will be possible only by leaving Earth for an outer planet long before the Sun starts to expand into a red giant.

Looking at several theories, human explorers would have to travel at least as far as the planets Uranus or Neptune in order to escape the red giant solar "heat wave." That is provided the atmosphere of both planets would not be adversely affected through increasing heat and ultraviolet radiation from the sun. As a matter of fact, the human "immigrants" would have to live in special enclosures in any case, to escape and survive the hostile atmosphere of either planet — as they are both loaded with deadly gases.

As referred to earlier, the density of the white dwarf "Sirius B" is about 100,000 times that of water, with a mass about equal to that of our sun, but with a diameter of only about two percent of our sun! Therefore we can assume, if a person or an object could be transported to Sirius B, it all would weigh 100,000 times the terrestrial weight. To make it more interesting, persons and objects would be only 1/100,000th their current height and dimensions as on our planet!

THE BUILDING BLOCKS OF THE UNIVERSE, SEEMINGLY THE SAME EVERYWHERE

At this point, I believe a quick reference to the facts of chemistry and physics is in order, to make understanding of the behavior of matter, space and time easier to comprehend.

How could an object on the surface of another heavenly body be only 1/100,000th of its size on Earth? How is it physically possible? — The readers must be all more or less familiar with the basic building blocks of the universe: ATOMS and MOLECULES. It is also known atoms are invisible at the unimaginable tiny dimension of only about one hundredth of one millionth of a centimeter in diameter. To illustrate it in another way, an atom magnified 100 million times would appear about one inch in diameter on a TV screen.

How can the tiniest component of matter be compressed to 100,000th of this already invisible dimension? The answer lies in the fact, that atoms are mostly "empty space," so to speak. The super-small core or "Nucleus" is the only solid part of the atom, — that is without going into details on electrons and protons, etc.

To illustrate the proportions of the nucleus compared to the entire atom, we can place an orange somewhere in the center of the Sky Dome Stadium in Toronto, Canada or in the Dome of the Capitol, in Washington, DC. The orange represents the nucleus, the rest of the space in the building is just that, "space" occupied by the entire atom.

To simplify the rest of the information on atoms further: The nucleus contains one or more positively charged "protons" and the same, or slightly higher number of "neutrons," surrounded by an equal number of negatively charged, orbiting "electrons." Incidentally, the existence of neutrons was not known until 1932. The most abundant element in the universe, hydrogen, is also the simplest and "lightest," it contains one proton and one electron.

Here are some other and better known elements, (the number indicates total electrons and protons per atom):

Helium He 2	Carbon C 6	Nitrogen N 7
Oxygen O 8	Sodium Na 11	Iron Fe 26
Aluminium Al 13	Nickel Ni 28	Lead Pb 82
Silver Ag 47	Gold Au 79	Platinum Pt 78
Silicon Si 14	Uranium U 92	Plutonium Pu 94

Although, the elements are not listed in any particular order, one can note a certain loose relationship between density and the number of electrons/protons. In other words, it is well known from earlier chapters that the elements hydrogen and helium are so light, Earth' gravity can not keep these gases near the surface. At earth-bound temperatures, hydrogen and helium remain in gaseous form.

Most hydrogen on Earth remained captured only by combining with oxygen in form of water molecules, (H_2O). The existence of helium was discovered on the sun first, before finding relatively small pockets of the gas on Earth also. Helium is the element, which may turn into carbon after going through a chain of nuclear fusion reactions, and after our sun will initially turn into a red giant and later into a white dwarf. Our sun may already have gone through two or three cycles of birth, main sequence and death previously, based on the existence of a relatively higher proportion of heavy elements.

The elements with the highest number of electrons/

protons are also the "heaviest," such as gold, lead, platinum, uranium and plutonium. Uranium for example packs 92 orbiting electrons around a nucleus containing 92 protons and 140 neutrons; therefore, this atom wastes less free "space" and is more densely packed. Plutonium, as one of the heaviest (man-made) elements, is also the most dangerous to living cells, due to its high emission of radioactivity from "alpha" particles.

Theoretically, it is possible with one pound of plutonium to cause lung cancer in every person on Earth. Fortunately, it is physically impossible to distribute the necessary minute quantities to five billion pairs of lungs to cause this ultimate demise. If a hypothetical "Satan" exists, could he achieve his goal of destruction, or if an "Almighty God" exists, would He then save us from this method of mass extinction?

Fortunately, real life is different. If anyone subjects himself knowingly or unknowingly to this type of radiation in sufficiently high dosage, he WILL be severely "burned" with fatal results. Prayers will not save him either. The people of Hiroshima and Nagasaki were the first such "victims of (tragic) circumstances," prayer and belief or non-belief in God did not save them from death in agony. Nobody and nothing escapes the laws of nature, the laws of the universe and the laws of physics, when it comes to the deadly effects of plutonium.

Hydrogen is not only the most abundant matter in the universe, closely followed by helium, it is also the "youngest" atomic matter, as it STARTS the chain of nuclear conversions through super-high temperatures etc., first into the next "heavier" element, helium. The same process continues on all the way down to elements such as the uranium atom, which is about 240 times as heavy as hydrogen.

It can therefore be safely assumed, the heavier elements are also the oldest and the rarest in the universe. According to this, some scientists estimate the universe is still relatively young; considering also, that most of the hydrogen present within the universe has not yet been accumulated in stellar objects, such as nebulae or stars where the conversion process starts. This brings me to the end of the short lecture on atoms and matter in the universe.

PROBING INTO OLD MYSTERIES OF DEEP SPACE WITH THE HUBBLE TELESCOPE

After the recent successful repair of the Hubble telescope, astronomers and other scientists are anxiously awaiting a deluge of new information. If I remember right, one religious sect stated, that "God did not want" the Hubble telescope to work, since it could reveal information contradictory to the Bible. Well, the doom and gloom prophets were wrong again. This masterpiece of modern technology IS now working seemingly perfect and the information coming from it, may in fact be revolutionary.

It is expected that some of the discoveries may force earlier scientific theories into obsolescence. In any case, the biblical myth of the creation (of the universe) will surely be more contradicted than ever before. It has to, since that version is already so far from the scientific truth that it takes very little to upset the biblical "theories" even further.

As mentioned, most scientists accept the "Big Bang" theory. Other amateur scientists, including myself prefer to be a bit more sceptical about it, as we would want to know what happened BEFORE the Big Bang. Because of its importance, I will try to explain in the most un-scientific, easily understandable manner some of the theories, findings and expectations. Chances are, by the time this book is in print, scientists will have accumulated a number of additional discoveries from the Hubble telescope, thereby making some of today's theories obsolete.

Considering the vast distances in the universe, 100 million light-years is not all that much, at least not compared to galaxies as distant as 10 billion light-years. Yet, we can only detect INDIVIDUAL exceptionally large objects (except quasars) with the best earthbound telescopes, up to 100 million light-years distant. Individual, even most massive stars can not be seen nearly as far. Astronomers will experience a major improvement over the current situation after the information from the Hubble telescope is all evaluated. We may be able to observe single objects in space, possibly up to one billion light-years distant.

What we will see then, has really happened when life on Earth was in its early stages and certainly long before primates and Homo sapiens evolved. The Hubble telescope will give us a much clearer visible, ultraviolet and infrared view of the most distant galaxies. The knowledge to be gained through this latest achievement in technology is endless. One could consider it to be the greatest advance by astronomers and scientists, since the invention of the first crude telescope by Galileo in 1609.

Not only will researchers be able to probe DEEPER into space, they will at the same time be able to look further into the PAST. We will find out more about quasars, the mystery objects of the past, so distant that their existence was unknown until about 1960. When first discovered, they were powerful sources of radio waves only and it was established that we were not dealing with distant star clusters or small galaxies. Until that time, quasars were nothing other than very faint, extremely distant light-sources.

In December 1993 with "near-sighted" fuzzy images received from the Hubble telescope, we have already picked up significant new information about the universe: *elliptical* galaxies are quite common in our interstellar neighborhood. After observing information from a cluster of 300 very distant galaxies, it became clear that *spiral* galaxies were more common five billion light-years ago. The question has now arisen: are elliptical galaxies originating from spiral ones? One would have expected it to be in reverse.

In the "Andromeda" galaxy at about two million light-years distance, a "double nucleus" was found, raising another question: did a smaller galaxy actually collide with the larger one and did the intruder get "absorbed" by the larger nucleus? The scale of the universe can soon be more accurately determined with new information through Hubble, from about two dozen newly discovered rare Cepheid variable stars.

The Hubble telescope already has provided science with new information about the actual existence of one of basically theoretical "Black Holes," within a massive elliptical galaxy at about 45 million light-years distance. At the core, through a frisbee-shaped void, gases and dust seem to disappear into darkness, into what appears to be a "Black Hole."

Another most significant discovery was that of pan-cake shaped disks of interstellar dust, around fifteen newly-formed stars in the Orion Nebula. This phenomenon could confirm the formation of planets around many solar systems.

QUASARS — ARE THEY PART OF THE ANSWER ON THE FORMATION OF THE UNIVERSE?

The reader may remember the "red shift" of light rays, caused by the same effect as the change in tone from the whistle of a receding train; Quasars are so distant from us, that the red shift is extremely noticeable. For example, the distance of one of the better known quasars, the 3C273, was calculated to be about 3.1 billion light-years distant from the solar system. At that time on Earth, primitive life had just evolved. The emission of light rays and radio waves from this object is about as powerful as forty of the largest galaxies. It appears ever more that Quasars are the breeding grounds for galaxy clusters. However, luminosity and radio wave measurements over a period of a few years have been found to fluctuate greatly, from a low point in 1964 to almost double the energy output around 1967 and back to a low point in 1979. Since then, energy measured from 3C273 has erratically increased again.

Some of the most distant galaxies in the vicinity of quasars have been calculated to recede from ours at the incredible speed of 200 million miles per hour, or at about one third of the speed of light. It is not likely that we will ever be able to visit THESE galaxies, no matter how sophisticated future space-travel will be.

The amounts of energy produced by most quasars is simply incomprehensible; just as another example, 3C273 produces as much energy in ONE SECOND as our Sun produces in ABOUT THREE MILLION YEARS! Yet the dimensions of quasars are "only" from about a few light hours "diameter," to little more than one light-year — almost comparable to the distance from our sun to the nearest star of about four light-years.

A quasar with a one light-year diameter may emit about as much total energy as 100 milky-way-sized galaxies, each of them about 100,000 light-years in diameter! Let us assume, all the visible stars (Northern and Southern hemisphere) were packed into an area of 100,000th of the entire space, now occupied by them. Now we multiply the total luminosity by an additional 100 times, giving us only about one third the light emission of a large quasar! This calculation is based on the fact that mankind can barely observe one third of our galaxy's total number of stars, and then only through powerful telescopes. Without the aid of a telescope, a terrestrial observer can barely count a few thousand out of approximately 400 billion stars within the Milky Way, exactly what our ancestors saw thousands of years ago before us.

Quasars are most unusual in behavior and increasing in their numbers, as we probe the much younger universe of the past to the limits. Quasars are the most distant, detectable and relatively "compact" objects. Their density and light emissions are simply of enormous proportions. Yet, at ten billion light-years distance quasars are visible only through the most powerful telescopes.

Each quasar is said to have an energy emission of as many as one hundred galaxies, but at dimensions close to one single "small" galaxy only. Could each quasar be the breeding ground for future super clusters? To look at it again, the energy emitted in one second by some quasars has been calculated to be about the same as our sun is radiating into space over a period of ten million years.

Some scientists believe that quasars are exploding (dying) galaxies, rather than nurseries for new ones. However, there is little logic to this theory, since quasars are found only at the greatest measurable or calculable distances in the universe, as they were occurring three to ten billion years ago. At that time, the universe was up to ten billion years younger, with little or no "old" and dying galaxies. Our own galaxy cluster is thought to be "only" about ten billion years old. At the measured age of quasars, it was possibly just about to evolve from a quasar ten billion years ago. That is, if it could be observed at the present time from a point in the universe, ten billion light-years distant.

As a matter of fact, it would be quite logical to conclude, that our local super cluster of galaxies appears as a quasar to a hypothetical observer ten billion light-years distant. It would have to be an extra powerful quasar, since our super cluster consists of thousands of galaxies. Or was it a number of smaller quasars? This alternative would be just as logical and acceptable.

Since there are no quasars anywhere near our region of the universe, scientists can only speculate WHAT they are, causing a number of theories to evolve. The most acceptable theory is the one of "breeding grounds" for galaxies. It would also explain the existence of such galaxy clusters, with all the member galaxies RECEDING from one another. However, there are other theories and we may soon find the answers with the help of the Hubble telescope.

BLACK HOLES, PERHAPS THE STRANGEST COSMIC PHENOMENA

Astronomers have calculated their existence, but the explanations for the formation of Black Holes are rather bizarre. Simplified, when stars reach the end of their life cycle, all nuclear fusion reaction is long gone and gravity is no longer counteracted by outward forces from the earlier nuclear furnace. The inward pressure from the collapsing heavier atomic matter now takes over.

As an example, what has developed into a "neutron star" with a total of about three solar masses to start with, will eventually collapse to the so-called "Schwarzschild" radius of less than six miles. Hard to imagine our own Sun in comparison, shrinking to a radius of less than two miles! A neutron star, similar to a black dwarf, is considerably more dense and much smaller than a white dwarf. The empty "space" around the nucleus of atoms has nearly all but disappeared during the process of ultimate collapse.

Just before total collapse, and as a result of ultra-high density and the massive gravitational field, the "photons" of light rays begin to bend. As the collapse and the density continues to increase, photons (light rays) are trapped by the immense gravity, their curved path is now pulled right back into the dead "star." At this point, any light rays still emitted from the object become invisible and a type of "Black hole" is created.

The activity can continue until, according to one of Einstein's theories, "singularity" is reached. At this time matter has squeezed itself so small, that it no longer occupies space! How can matter be so unbelievably dense and still exist, but at the same time not occupy any space? According to Einstein, it seems to be possible, and he has been right before.

The more comprehensible end to the life of a very large star at the point of highest density, is a super nova explosion, when really tremendous amounts of energy are released. The formation of supernovae was brought up in earlier chapters and at this time, I want to add only that supernovae do not appear to end in black holes.

This is where humanity still has to find the solutions, just as we had to find so many answers through science before about the mysteries of the universe. Another theory of creation of the universe is based on the same idea, when at the "outer limits" of the universe, ALL matter converges into black holes. Black holes also appear to be a reality at the center of some galaxies, where the density of mass becomes so dramatic that the inward rush towards a relatively tiny center forces matter to disappear into a black hole. Like a liquid is sucked through the bottom of a funnel before it disappears.

It is thought, the "disappearing" matter re-emerges eventually again on the other side of the black hole in form of "jets," where the material is dissipated into intergalactic space, beginning to form new nebulae and small galaxies.

The reader may remember the resemblance of the "vortex" action of a hurricane on Earth, compared to that of a spiral galaxy, so much that photographs of both natural forces look astonishingly similar. One can possibly compare the eye of a hurricane and its highly pressurized dense center, with the black hole in the center of a galaxy.

MYTHICAL RELIGIOUS "CREATION" VS. COSMIC CREATION

The Hubble telescope will expand humanity's "horizon" of the visible cosmos from the current limits. It may possibly alter our theories about the limits of the theoretical universe, back to the time of the so-called "Big Bang." As mentioned earlier, most scientists accept the Big Bang theory as the creation of the Universe, at between ten and fifteen billion years ago. However, everybody will admit that this theory still has a number of loopholes.

According to Einstein's and Hubble's theories, time, mass and energy are combined in several scientific formulas, which lead to the Big Bang about 10 to as much as 20 billion years ago. At that time all matter and energy was created in form of "prototypes," followed by a long chain of events which lead to today's known universe. All calculations lead to this point. Except, this is when the theory becomes a mystery, time "zero." With the universe at "zero" dimensions, and the temperature at "infinity," when it was all to have started.

No one can imagine "zero" time or space and "infinite heat," and all that at the same "time." Supporters of the Big Bang also caution, one should NOT imagine it to have happened in the "centre" of the universe, since there WAS NO universe and consequently no "centre" before the Big Bang happened. That is even more impossible to understand; no space, no mass, no temperature, no time, no centre of space. Then came the Big Bang and so on.

There HAD to be something before the Big Bang. To make it more complicated, this theory also implies more or less that the universe HAS a limit, and that it really is "finite," since the theoretical time limit is up to twenty billion years. Therefore, anything beyond twenty billion light-years distance is non-existent, or possibly the "curvature of space" leads us right back to the present?

I tend to lean more towards an "infinite," pulsating and curved-space universe, i.e. a Big Bang was the result of contraction of the mass and energy from a previous universe. When a tremendous pressure build-up and dramatically increasing heat built-up caused the Big Bang, the reconversion of matter and energy through various stages of nuclear fusion and gravity was triggered again, to the present stage of a "young," hydrogen-dominated universe.

Some evidence seems to point to this theory, re-enforced by the existence of only very small proportions of a few "old," HEAVY elements, possibly carried forward from the previous universe? It becomes logical, since stars are constantly created from interstellar dust, which originated from material of dying and dead stars in the first place.

Eventually, the expansion of the present universe will

slow down more and more to the point of reversal, and to the start of renewed contraction until the next Big Bang. Even that idea is still left with one point: How can space be of infinite dimensions, how can *anything* be infinite? — But at least, the thought of "No time, no universe, no time, no temperature" before the Big Bang is eliminated, science is left with only ONE "impossibility."

As mentioned, scientists have a few other theories about the creation of the universe. Following is a short list of a few of them: One idea is the existence of a "parallel" universe or several of them all at varies ages. This alternative has been combined with "anti-matter" universes, which shrink while ours expands and visa versa. The existence of anti-matter has actually been theoretically calculated, it was to have played an important role during the earlier stages of the creation of our "present" universe.

Current research continues to bear out the validity of parallel universes and their interconnection through so-called "wormholes," all part of a "hyperspace" theory. Somewhat logical, but as yet un-proven research puts our universe into an endless colony of parallel universes, connected through wormholes. As one universe expands, it finally disappears through a wormhole into another universe, creating an even greater vacuum in the original space. Another universe expands into the newly created space and appears through another wormhole as a "Big Bang" — and so on.

Similar to the theory on parallel universes is the idea that each living organism on Earth is a "micro universe." Taking the human body as an example with about 75% water content, each water molecule with its two hydrogen and one oxygen atom may be looked upon as a "triple star system" with a number of "planets" orbiting around. "Stars" are the nuclei, electrons and protons the orbiting "planets" with a lot of "space" in between.

Mixed in with singular atoms — such as calcium — to the most complex protein molecules, each body organ would represent a sort of "galaxy" of billions of stars and the entire body ("expanding" from conception and during growth) becomes a micro universe.

Over five billion such human micro universes exist now and their countless trillions of water molecule "star systems" are constantly re-cycled (or re-distributed after death) with others. The life span of all individual "star systems" will only end when the sun becomes a red giant in about four to five billion years from now. The point is, there are many ways to interpret the known facts and theories and this is just one explanation for the possibility of an existence of parallel universes. In either case, the numbers of "solar systems" are unimaginably vast and incomprehensible.

It sounds all a bit like science fiction and very incomprehensible; considering that even ONE universe with billions of galaxies, each with billions of stars, planets and moons of its own is seemingly endless at a diameter of say, fifteen billion light-years. How insignificant is our own tiny speck, planet Earth, the biblical "centre" of the universe, compared to ONE known universe alone. How about an endless colony of parallel universes?

Yet another idea is based on the constant "increase in receding speed" of the most distant, visible galaxies and quasars. This increase in acceleration, combined with the "red shift" cannot exceed the speed of light. Even if matter COULD accelerate beyond the speed of light, we would be unable to see it from our position in space. However, if it really is possible for matter to exceed the speed of light, (contrary to most current theories), matter would disappear into "Black holes."

Religious believers certainly cannot find any answers at all in the old holy books and scrolls for most of what astronomers have discovered just recently. Fundamentalists have given themselves little choice, other than to continue believing in the myth of biblical creation and in the usually contradicting legends from various major religions. Maybe until something better comes along through another new and different interpretation of a scripture?

They are denying themselves a choice between a sceptical re-evaluation of their ancient teachings and the known facts and common-sense-theories of science.

Maybe a more realistic approach can be expected through the eventual contact with highly advanced extraterrestrial visitors. They may just have most of the answers for us earthlings? That is, if they want to share them with us — maybe when Homo Sapiens finally displays obvious signs of common sense including an end to primitive behavior through religious and territorial wars.

All things considered, the Cosmos may not have required the act of "creation," if one can accept the incomprehensible theory of an infinite Universe. *What has always been here, had no need to be created.* Matter converts into energy and back to matter, and one element converts into another in one of billions of a nuclear fusion "furnaces," and without an actual loss of matter or energy ever. It makes a bit more sense than biblical "creation."

Even if the Big Bang theory will be proven to be correct, without the existence of a previous universe, the "evolutionary changes" of the universe so far can not really be called "creation." In any case, the standard version of the Big Bang theory is not really as acceptable as the "cycling" or "pulsating" version, where the Big Bang repeats itself all over again.

The reader had an opportunity to evaluate the physical impossibilities for a single "Maker," "creating" every star and all planets, to give and to take each life on Earth and everywhere else (outside this planet, if one accepts extraterrestrial life), and at all times. The sheer numbers make it impossible for even the most superior, hypothetical being: the insurmountable astronomical distances and natural events taking place in form of terrestrial and interstellar evolution, in destruction and re-forming of planets and stars — all making a single "creator" at best hypothetical, ancient science fiction.

The laws of physics and nature and evolution are identical, wherever they can be observed or measured in the universe. We have compared the close resemblance of all life forms, including their similarity in "creation," life and death.

The reader had the opportunity to re-evaluate the physical impossibility of all, or some of religious believers' prayers being heard by an "Almighty" God. And all of it taking place at the same time, when five human beings are conceived, born or die every second on this one planet alone. We have looked at evolution, how it affects ALL life-forms in accordance to changes in the environment and how much nature is constantly trying to catch up to those changes.

Maybe the physical impossibilities of a "God" being present everywhere, and controlling several hundred thousand life and death events at the same time, made formerly blind "believers" take a second look. Re-evaluation possibly may leave them with no longer accepting a "God" in human images, they may begin to think of the universe itself as the only real "God"?

If firm religious believers want to re-name the Cosmos to "God" or God to "Cosmos," many non-believers can probably accept this basic name change for what it is worth. But by doing so, one must question "Jonah" 10:4, 10:6 in the Koran: "It was He [Allah] that gave the sun his brightness and the moon her light, ordaining her phases that you may learn to compute the seasons and the years. God created them only to manifest the truth." In other words, if God IS the cosmos, he cannot have created it, or not JUST the sun and the moon.

Nothing in the universe creates itself from nothing, with the exception maybe for those, who firmly believe in the Big Bang where all matter in the universe started from nothing, without time or space. The reality is that even the Big Bang theory is strictly referring to "time, space and matter," NOT to a mythical God, who is supposedly everywhere at all times and who can "help" and "love" us and who decides over all life and death.

The universe, as it is known at the end of the twentieth century, also can not "hear" prayers, it cannot accept prayers of thanks for food or for good health, this universe is not a human-like figure with ears and eyes. Man is not "created" in the universe' image, we are part of it, we are inside of it, all part of a global community. Our bodies are made from the same atomic matter, which is all around us and which can be observed or measured millions of light-years away from us. And it is also fact that during the process of evolution, the same forces, which create life also take it away.

Our sun is the basic REAL and immediate source of life on Earth, yet it will destroy all of it, when this insignificant star out of billions eventually dies itself. A new sun may again form out of the remnants of the old one, mixed with additional debris from other extra-celestial sources. Billions of years from now, there may very well be a new "Earth" and human races may evolve on it again, but the conditions will never be the same. As stated, no two trees are the same, no two people are really identical, and there are no two stars or galaxies totally identical . . .

A hypothetical, new and "re-cycled" planet Earth may just be slightly, or considerably different from ours, enough to enhance the development of intelligent being more and faster, or not at all. I can accept most scientific theories about the creation of the universe and of life, as long as they are based on scientific fact and logic. But I cannot accept the ancient writings of one religion or another, one out of over 150 or so, when they are NOT based on logic or scientific knowledge. Equally, I cannot accept the existence of a biblical "Heaven" or a "Hell," unless life during a senseless, vicious military conflict in one country or another (Rwanda, Bosnia) is called "living hell."

The last section is coming up and dealing with how our planet COULD look like in just a few hundred years from now.

SECTION III

THE FUTURE — WILL IT BE GUIDED BY KNOWLEDGE AND LOGIC AND LESS BY RELIGIOUS MYTH?

UNCONTROLLED WEATHER EXTREMES: ONE OF THE MOST DESTRUCTIVE FORCES ON EARTH

Whenever, (or "if," depending on one's belief) Earth is visited by highly advanced extraterrestrials, their first impressions of earthlings and the condition of the planet would have to be based on logic. They would have to immediately notice the failures of humanity. It would be obvious to them how little improvement to life on Earth has been accomplished over the past 5000 years, with or without the "help" of man-made, mythical religions, which too often were humanity's priority. An intelligent and technologically advanced visitor would probably immediately notice humanity's illogic in priorities. The visitors would soon discover the major problems on Earth and their causes. I want to list some priorities, as an extraterrestrial visitor might just see them.

I have simply asked myself: What is the most obvious cause of death and destruction through a number of particular natural disasters, aside from man-made armed conflicts? The answer is simple: WEATHER EXTREMES!

From the fires in California and Australia to the floods in the Mississippi Delta and California, to nearly annual floods in Bangladesh, from death through malnutrition in Ethiopia and in the Sudan, to death and destruction through blizzards in North America and in Europe, to earth-scorching, devastating heat-waves in many other parts of the world — The WEATHER is the principal cause of it all, without question. And what is humanity doing about it?

Many controversial books and articles have been written about the subject of weather control, the idea is certainly not new. However, since this project would be so immense and long-term, aside from unimaginable cost, it has been rejected so far and pooh-poohed in some cases as totally utopian. However, electricity, horse-less carriages and telephones were also rejected as absurd not so long ago, and the first experimental "flying machines" were laughed at.

Of course, there may be seemingly insurmountable obstacles at first glance and a delicate balance would have to be maintained. Too much man-made "adjusting" in one area would cause havoc in another part of the planet. In my opinion, the answer is in detecting the very embryonic sources of storms.

It is no secret, major storms ARE caused by dramatic temperature changes. So, the next question is: what causes drastic temperature fluctuations? For example the sun's heat only decreases very gradually during the hurricane season in the southern United States. Why then DO hurricanes and tornadoes form rather rapidly and why do they form in THAT particular area? Why did Hurricane Andrew devastate a particular part of southern Florida, why not elsewhere where previous hurricanes did their job years ago?

What comes first, the storm or the temperature change, and why is it happening every year in different areas of the same, sometimes vast regions? Why are there sometimes rapid temperature changes WITHOUT causing violent storms? Why are there rapid temperature changes at all? Not being a climatologist or meteorologist, there may be SOME answers out there already to my questions. But the basic question has not been answered yet: where does it all start, what tiny cause triggers a major storm and what can we do to eventually stop it, long before it grows from a "micro"- whirlwind into a devastating storm? The energy released by ANY strong wind is awesome, when one considers how millions of square miles of lake and ocean surfaces are nearly constantly kept in motion. And it goes on nearly year-round in most parts of the planet.

The "jet stream" — an upper atmospheric wind — in the northern hemisphere has a most dramatic effect on the climate of the entire northern part of the United States and all of Canada. Most frequently, what causes unusual mild spells or periods of deep freeze during the winters can be attributed to a shift of the jet stream. Although it is always present, it occasionally loops far to the north, causing a pleasant mild spell to the south of it, or it shifts far to the south, resulting in frost in the southern states.

The jet stream's vast shift in either direction must start somewhere like everything else in nature, from an embryonic stage. Will scientists be able to eventually divert the jet stream during the winter to the north from this early beginning and keep it there all winter? The thought of regular mild winters in the populated north looks very enticing — let's think of the savings in fuel, the reduction of accidents caused by snow and ice and so on.

Annual hurricanes and typhoons cause massive destruction and loss of life, human and animal. While certain parts of the world are repeatedly trenched by excessive rainfall, elsewhere millions of square miles of desert have been useless wasteland for centuries. Great parts of northern Canada and Russia are equally wasted because of weather extremes. Plant life could thrive, given the right climatic conditions, even in parts of Greenland and Antarctica, while the excess water from a slow reduction polar Ice caps could be utilized for irrigation of desert and other scorched areas. Only 19% of the land surface of Earth are inhabited.

While humanity was able to spent about 800 billion dollars annually on weapons of destruction during the cold-war-years, most of this massive amount of money could have been "invested" into research and development of "weather control" installations. Once scientists know exactly why and how a hurricane or tornado really starts, from a very insignificant embryonic beginning, the means of prevention can be developed also. Maybe a fraction of the awesome and seemingly unlimited and wasted wind power could possibly be harnessed for useful purposes, given enough time and funding for research and development in the first place.

The forces of "just" a strong wind are simply striking, when massive wave-action can develop relatively quickly over large bodies of water. There is no man-made power source so great and powerful, which is able to stir up one of the Great Lakes, or sometimes all of them at the same time, like gale-force winds can. At this time, I am not even considering the wave action of the vast oceans of the world. When will mankind be able to harness only some of this force, thereby reducing the devastation from MAJOR storms and the resulting wave action along shorelines?

To go one step further, the savings of conventional energy (and subsequent environmental pollution), would be tremendous. If climatologists and other researchers were able to initiate a slight change only to temperature extremes in so-called moderate zones of the planet, the results would be astronomical. A shorter winter in Canada, the Northern US and Russia, the elimination of severe cold spells and of lengthy heat waves, increased rainfall in desert areas and less rainfall in parts of the world, where there is consistently too much of it.

The savings would pay for the cost of hypothetical weather control installations in a relatively short time period. The changes would have to be very gradual, along with other precautionary measures, to allow all of nature to evolve accordingly, giving evolution time to work at an acceptable pace.

Mankind found funding and had enough scientific knowledge to send men to the moon and to design and manufacture missiles capable of exploding a thermo-nuclear warhead with pin-point accuracy on a target 6000 miles away. We should therefore have little difficulty to concentrate on research and development of weather control methods, for the BENEFIT of all mankind, if we put our mind to it. Controlled weather IS really a number "one" priority, it is no longer utopian.

Ultimately, hurricanes and typhoons would be eliminated, the winters would be shorter and milder, lush green forests would grow in former "deserts". Annual floods and gale-force storms driving the oceans' waters into the shorelines would be a thing of the past. It may sound like fantasy. But the time has come to think seriously about it: "PREVENTING" instead of "CLEANING UP." Yes, the oceans may gradually rise a meter or so, but with lesser wave action, the higher water level would be quite harmless, acceptable and beneficial in some areas.

GRADUAL ELIMINATION OF INTERNATIONAL BORDERS TO COMMERCE, ULTIMATELY A WORLD GOVERNMENT

A costly and often fatal tradition of humanity can easily be identified in the existence of artificial, sometimes ancient and usually senseless national borders. Hypothetical highly advanced alien visitors would in all probability not be able to comprehend the existence of such borders, when throughout the history of humanity the same borders directly caused most military conflicts.

After "Free Trade Agreements" have already been initiated, a controlled and gradual opening of national borders would eliminate the reasons and possibilities for military conflicts. It would also overcome ancient prejudices and inherited hatred for people of different religion, language and culture. Most European nations have already been successful with their newly formed relationships, far beyond just "free trade." Former national borders could be gradually turned into "provincial," "territorial" or "regional" borders, all under the general democratic rules of a (reformed) United Nations government.

The best example of an existing similar political systems, aside from the European Economic Community, are the United States of America. There is no longer any reason for military conflicts over national borders between various U.S. states, no more than between the various European member countries of the EEC. It is most remarkable for Europe, since the EEC member countries had been involved in major armed conflicts against each other, twice during the twentieth century alone — the last time only fifty years ago.

While the former Soviet Union had at least managed to keep fifteen individual states working together in relative harmony for nearly seventy years, the former republics have now each gone their separate way and backwards in history. By separating the former union, various armed conflicts between former reasonably friendly republics are brewing or are in progress. Yugoslavia is an even worse example.

It is beyond most intelligent observers why instead of trying to re-built their shattered economies first, their priorities were in DIVIDING an existing entity. An uneasy relationship between the various states of a union is still much preferable to no union at all and to a senseless military conflict over artificial borders. The devastating results to all involved are only too obvious, aside from a tremendous financial strain to a number of UN member nations. Over 200,000 former neighbors have killed each other so far in the former Yugoslavia for inexplicable, illogical reasons. The devastated towns and properties will take generations to re-construct for winners and losers alike. And the lost lives can never be replaced.

Although humanity may not be ready for a world government yet, this solution should be a goal for all nations. Using the example of the European Economic Communities or the United States of America, all nations could ultimately remove trade and population borders first, at the same time eliminating the need for massive defence budgets. Further down the line, the world community could be governed by a system similar to that of the United States, where individual States still govern their own territory to a point. A world, where ultimately one currency and free movement of goods and population could be a reality in less than fifty years.

A world government would consist of elected representatives from each former nation, somewhat in proportion to their individual population; i.e., the "former" China and Russia would have more representatives than Liechtenstein, etc. Much has been written about a world government, but it is really the ONLY logical alternative to the status quo. The past and present situation is not working and it never has — the killing of millions during territorial conflicts has gone on too long, next to killing of religious "victims of circumstances."

A true democratic world government CAN ONLY be established when mankind is ready to give up radical nationalism and certain senseless traditions, without giving up most cultural heritage. Amongst other obvious advantages, food production could be centrally controlled, along with the environment. One world currency would eliminate exchange problems. Birth control could be mandatory for former nations with a stagnant economy to allow the new "state" (or province) to catch up to their "neighbors" somewhat quicker. Unfortunately, it may sound like an utopian dream right now, but it is possible and it may just all happen in the next few hundred years. Who would have thought, the EEC would ever become reality, only fifty years ago. I think, it WILL have to happen.

ANOTHER CHOICE: THE ANNUAL LOSS OF TENS OF THOUSANDS OF LIVES OR GIVING UP "THE RIGHT TO BEAR ARMS"?

A change of most religious or, in this case, tradition-based laws is an absolute necessity, based on pure logic. Modern laws strictly control most dangerous sports, such as automobile and boat racing, by allowing them in designated areas only. As a result of it, a few lives are saved. In other words, if auto races were to be allowed on unprotected highways, other innocent drivers and passengers would surely be killed.

Obviously, automobile races had to be prohibited on public roads and legal speed limits are enforced to prevent fatal car accidents, (although some limits are ridiculously low). In addition, many states and provinces have seat-belt laws, again saving lives. At the same time, we allow basically ANYONE from age 16 years up or so (depending on local laws), to purchase a much more dangerous piece of so-called "sporting equipment" — a HAND GUN or an assault weapon. After loosing between 15,000 and 20,000 lives annually in North America to gun-related crimes and accidents, how long will it take to stop this senseless massacre in the name of "sport" or "protection" (from what)?

Considering the number of lost lives from gun-related causes, what does it take to have the law makers introduce gun-prohibition? Another 25,000, 50,000 or one hundred thousand lives? The statistics alone should cause every lawmaker to act on immediate changes: U.S. gun owners kill themselves or family members forty-three times as often as they kill attackers, about 40% of all gun-related killings are by the owners' own weapon.

It is so easy to pull a trigger. In the U.S. alone, over 200 million guns are in circulation. Of them, 60 million are handguns. Death by gunshot wounds is second only to fatal automobile accidents for children of high-school age. On average, twenty-seven Americans die daily from handguns fired by others and thirty-three commit suicide daily with same kind of weapon. During 1990, 64% of a total of 23,438 murders committed involved firearms, a round 15,000 gun-related deaths. In three years as many as the total of American lives lost during the entire Vietnam war! Am I missing something? To top it off, President Clinton's relatively "mild" Bill on crime was rejected by Congress at first and had to be re-written!

There could be a number of logical ways to accomplish and enforce complete gun prohibition, if only to save a few thousand lives during the first few years. The results would be most encouraging when a few thousand lives ARE saved, the initial battle is won! One way of effective gun prohibition would be in the beginning a very thorough test and a much higher minimum legal age for gun buyers. The membership in a gun club should be compulsory. The owner of a gun would then only be allowed to take the weapon to the club, where it would be locked up permanent and where it would not be allowed to be removed from the premises. Anyone caught with a gun outside a the registered gun club would be subject to confiscations of property, stiff fines and prison, and no matter who it is.

In comparison, a recently released statistic stirred considerable controversy and several heated public discussions. The subject discussed were the relatively new "in-line skates," roller skates with four wheels in line rather than two pairs of wheels side-by-side. Some considered them extremely dangerous and even suggested a total ban. Besides a number of minor and major injuries inflicted to mostly children, there were six fatalities in one year. Nearly all of the fatal injuries were caused by carelessness on the part of the victims.

Children will be children and just as they occasionally jump in front of a car while running, they will do the same on roller skates. The point is once again: SIX fatalities start a major discussion and a call for the ban of an in itself innocent piece of sporting equipment, while 10,000 or 15,000 annual victims of shootings do not cause the public to demand prohibition of the not so innocent instruments of deaths and injury. Where is the logic? When will the pro-gun lobby groups look at the plain facts and admit that guns ARE designed to kill and not just for "harmless target practice" or hunting?

Possession of a small amount of marijuana in some countries or states is a major offense, yet possession of a potentially deadly weapon is legal in the same places! Drivers are fined for not wearing seat-belts, but they could carry a deadly weapon in the car at any time, as long as it was bought legally! As I previously commented, in Tokyo, a city of over 10,000,000, only seven or eight people are killed annually through gunshots — how many in Washington, DC — 800 or 900? How about Miami, New York or Los Angeles?

Before gun prohibition would become law, the authorities may have to buy back guns during an amnesty period. The cost of a gun-buy-back program would be minute compared to legal costs, victim's medical expenses, law enforcement and prison maintenance costs, currently as a result of gun-related crimes. Prevention is always cheaper than cure, and in this case, it saves lives. Legal transport of guns could be restricted to police officers, security guards and the military, and only while on duty. Licensed hunters would carry a weapon only while travelling to and from a designated hunting district. All other gun possessions on the streets would be automatically a major criminal offense, subject to the most severe penalties and confiscations.

Again, if we can save a few thousand lives during the initial years of prohibition, and a few billion dollars in expenses caused by gun-related crimes, the first round would be won. The fact is obvious, the way we are carrying on NOW based on the traditional and totally obsolete "right to bear arms," is NOT working at all, and it will not work any better in the future.

Some gun owners in crime infested cities purchase hand guns for so-called protection. However, unless the owner carries the weapon at all times, is ready to fire in a split second and the would-be assailant approached from the front, the gun is of little use. Unfortunately, most gun related assaults are not from the front, but more often from behind or either side and in total surprise to the victim. As a matter of fact, gun ownership only adds to the probability of an accident, killing or injury by the owner's own weapon. During break-ins, a gun owner's weapon is often the prime target of theft by the intruders.

Of course, one may be tempted to compare the prohibition of alcohol during the thirties in the USA with prohibition of hand guns. It did not work, and alcoholic beverages can not be compared with a deadly weapon. Crime was actually CREATED with the introduction of the religiously biased alcohol prohibition, in the same way as laws are still enforced to turn any non-violent, victim-less act into a crime. Supervised, regulated prostitution, possession of marijuana, etc. are only two examples.

Although deaths caused by gun violence may not be on the top a list of a futuristic view of this planet's next millennia, it could be one of the easiest changes for the better. In fact, it would be much easier to achieve than "weather control" — Who knows, hypothetical alien visitors in a few hundred years from now may even see a society on Earth, where ALL guns and weapons are eliminated and no longer necessary, not even by the military.

It is statistically proven today, wherever civilians possess the greatest number of hand guns, the most gun-related crimes are also committed in that part of the world. Also, when more military hardware is produced by one nation, accordingly more weaponry is required by other nations in "preparation" for possible defence. In contrast, when weapons of destruction are eliminated and destroyed all over the world, and when the United Nations make production of all weapons illegal, military conflicts are automatically and gradually eliminated, along with criminal, gun-related activities by civilians. It is pretty difficult to fight a war or to commit a major crime with knifes only.

AFTER WEATHER CONTROL, LET SCIENTISTS "CUT OUT THE MIDDLE MAN" IN THE FOOD CHAIN

It takes literally tons of feed to raise a calf to the "age of slaughter." A cow goes through one of the most inefficient processes of nature — converting certain plant life into protein (meat) for human consumption. The human organism in turn operates in the same cruel and wasteful way as our carnivorous "brothers" in nature: We convert large amounts of meat, aside from vegetables and fruits, into new body cells and tissue.

The time may come soon, when scientists have to "create" meat in the laboratory from the most abundant, fastest growing and primitive plant sources. Over the centuries of human history, mankind has been taking the detour via living "laboratories" of nature, called cows, chicken, sheep or pigs for replenishment of body cells. Aside from elimination of the cruelty of slaughter of other living creatures as a food source for us, "laboratory meat" for human consumption would free vast agricultural areas for more "direct" vegetable food sources for humans.

Initial steps away from meat have already been taken, when soya-meal meat substitutes had been introduced several years ago. However, the creation of laboratory meat would be the final and more direct approach. In the meantime, the introduction of meat substitutes still requires considerable work and persuasion, necessary to perfect better substitutes and to convince people to accept them as a fully acceptable substitute.

Scientists have only recently discovered how harmful and significant the amount of air pollution is from livestock waste to this planet's environment. Massive amounts of methane gas and carbon dioxide are released into the atmosphere. In many parts of the world, the supply of water for life stock AND for irrigation of feed vegetation already causes serious water supply shortages. It is in those areas, where the introduction of meat substitutes and laboratory meat should be an important priority, while industrialized countries should follow suit shortly thereafter.

Over the next one hundred years, scientists may really be able to develop the more efficient sources of meat for human consumption through years of careful genetic engineering. Presently, the inefficient cycle in our food supply, where tons of vegetation are required for each pound of beef, is simply illogical on a planet with 5 1/2 billion inhabitants. And let's not forget the increased air pollution caused by animal waste.

Over-fishing during past years has resulted in a recent crisis to the supplies of certain fish species, normally abundant in the western Atlantic. The examples go on, all proof that the status quo cannot be accepted any longer. To start with, fish supplies should be treated in the same manner as de-forestation by the lumber industry and paper mills. Just as re-forestation is necessary to replace the cut trees, immediate international breeding programs for replenishment of the fish supplies should also be initiated, along with the already prevailing restrictions on ocean fishing.

It is difficult to understand, why our governments have waited so long to come to the conclusion that fish breeding programs must be encouraged. It has been considered more seriously only very recently, after it is almost too late for some fish species. Ultimately in centuries from now, humanity may have developed not only "laboratory meat"— sources, genetically engineered "fish" as a food source could replace our present day supply of seafood and fish.

So far, I have only picked some of the most obvious priorities for humanity, if the present sometimes chaotic conditions are to be improved. Or what our hypothetical alien visitors would immediately notice as the greatest flaws in human civilization. A number of other important changes are actually in the making or under consideration now, except not enough priority is given. I will select a few more possible priorities for present-day research.

MORE SMALL PRIORITIES, OFTEN OVERLOOKED

Throughout this book, I have referred to illogical regional and federal laws, mostly traditional and often religious-based. Many laws of various US states or Canadian provinces vary to a great extent from one another, causing unnecessary court actions, or duplications. A simple philosophy could be: if it works well in one state, why is it not acceptable in another? If there is no logical answer, why not change to the better system of the two? Lawmakers are known to debate for years over a possible change to an existing law or the introduction of a new one, while the law being debated is already successfully applied for years in another nation or state.

To name a few examples: regional environmental laws often differ not only between states or provinces, they are different from the federal laws. Because of the variety in environmental regulations, the cost to taxpayers only adds to the financial burden. If a substance is proven to be really harmful to life in one province or state, it must be harmful to residents of another. Or if a certain limit to emission of polluting substances is definitely found not to be harmful in one state, why is it not acceptable in another?

One basic federal legal system would eliminate the duplications and triplications. To be fair, a national referendum may be necessary in some cases to get a matter settled for once and for all. There may be exceptions to this idea, simply because of the variations in ecology from one state to another. In other words, alligators do not have to be a protected species in North Dakota while they, and manatees must be protected in Florida.

The penal codes are as different as the climate in various states: why is the (ancient) death penalty for first degree murder still in force in one state and only life imprisonment in another? Murder is murder, why is a convicted murderer executed in Florida and not in New

York until now? Only a federal decision could change this variance, and only after a national referendum, to establish if the death penalty is really at all acceptable.

In the opinion of many, including my own, the death penalty is too much religiously influenced in the first place. And as mentioned earlier, if one out of a hundred death-row inmates is executed first and found to be innocent AFTER, it is too late! Hard labor and “Boot” camps for convicts would be a better solution in place of some of the high cost hotel-type prisons as one extreme and executions as the other. Conversions of abandoned military bases into prisons is another good idea.

PROVEN premeditated first degree murder and PROVEN murder through terrorist actions against innocent bystanders could be punished by incarceration for life and by subjecting the offender to testing for medical research. Sounds maybe harsh, but rather than execution and disposal of the body, the offender could end up by actually saving human lives.

In general, representatives of various states, provinces and other nations should compare notes with each other and change all laws for the better. May it be immigration laws, anti-drug laws, education etc., etc. Scientists of various nations are doing it now, working together as teams on various research projects, rather than working independently on the same idea. It is the only economical way to proceed with new science venture.

National or state health insurance programs fall into the same category: why not take a look at the Canadian health insurance program and copy the better ideas ONLY, along with those of the German health care system and that of the United Kingdom, and visa versa, since neither system is quite perfect? Common sense is the only solution, all nations must learn from one another.

Earlier chapters dealt with a number of criminal and civil laws, which are solely based on illogical religious belief. A complete overhaul of the legal system would have to include a thorough examination of the penal system and length of incarceration of a convict. The current system is not working, almost any change would result in improvement.

At this point, I want to look again for "logic" in building more high-cost prisons, where convicts often live better, all expenses paid, compared to when they were on the "outside." At the same time, underprivileged law-abiding citizens live under most unacceptable, often inhumane conditions in nursing homes, etc., due to cut-backs in public funding.

A prison for violent offenders should NOT be nearly resembling a retirement resort, after all it is a PRISON. One solution for punishment and teaching discipline to so-called "young offenders" (except killers) has been suggested: boot camps. Reverting back to supervised corporal punishment (the strap), may be another most effective treatment for young vandals, rather than prison time. And after receiving punishment, the offender should be ordered to contribute towards restitution. Maybe the laws in Singapore are not so far out after all, it seems to work, youth vandalism is way down compared to North American standards.

Much to my amazement during the early fifties, in the Canadian province of Ontario, lashing was still used as punishment for a number of offenses, including vandalism. The results were remarkable, acts of vandalism were at a minimum. Similar crimes are severely punished in many Islamic countries. Although, I do not agree with the definition of so-called "crimes" in one country and not in another, and with the methods of punishment, the crime rate is extremely low in most of these nations. Theft by repeat offender is often still punished by amputation of the right hand. I am not suggesting this method for North America, but the strap or other corporal punishment would probably work wonders with young offenders.

It may not be so much the inflicted pain of the lashes, the humiliation seems to be a much greater deterrent. In any case, when I visited some of the most fundamentalist Islamic countries, (Libya, Saudi Arabia, Iran) I felt safer than I would in many North American cities. Corporal punishment, or the fear of it, DOES seem to work.

At the present time, there are still vastly opposing extremes between the penal systems of Islamic fundamentalist nations and the methods of punishment for crimes in western nations. Where in our (western) system the punishment for crime (other than the death sentence), is often too mild, punishment in Islamic and East Asian nations is often too severe for sometimes victimless "crimes." Such as their penalties for alcohol consumption.

When the young American in Singapore was punished for vandalism by four strokes with a heavy bamboo cane, (administered by a marshal arts expert), the punishment in North America for his crime would have been unquestionably too mild. It probably would have been a very short prison sentence, a fine or probation, with little or no "deterrent after-effect." The answer must be somewhere in the "middle:" Controlled lashes for vandalism, yes, but let it not lead to exaggeration either way.

Many former military bases have been closed after the Iron curtain was lifted and could be converted into "boot camps" for "young offenders" and other convicts, as long as they are not candidates for maximum security and do not represent a danger to other convicts. Another solution, particularly for more serious repeat offenders, would be labor camps in isolated areas, such as in far northern parts of Canada, or in the desert or other remote areas of various states. The profits from labor camps should than be made available to victims and families of victims of violent crimes. The point here is: the system, as it is "functioning" now does not work any longer, we can see the results daily. A dramatic change IS A PRIORITY.

Getting back to modern science; Another item of priority is a much greater effort for development of alternate energy sources, causing little or no pollution. Right at this time, with many defense contractors looking for business and sending people to the unemployment lines, why not concentrate on the development and manufacturing of more efficient solar and wind-energy sources? A few thousand wind generators are already in use, but the sky is the limit for any part of the country, where solar radiation is scarce, and where wind power is more in abundance. A small portion of the former defense budgets would do wonders.

By far too little attention is given to hydrogen gas as an energy source for automobiles or for that matter, as replacement of any petroleum-product-driven engine. Every once in a while, one can read about a new invention where hydrogen gas is actually used as power source for some sort of engine, but little is done to put serious and combined effort into it. One thing is for certain: conventional fuels WILL run out sooner than we can imagine, the present glut of crude oil supplies WILL come to an end. The question is: are industrialized nations prepared for it?

Here is an interesting foot note: The Japanese car manufacturer "Mazda" has just completed development of the first experimental hydrogen-powered car, the exhaust is nothing other than pure water vapor!

The next step into the future could be a massive, international scientific research effort to utilize the much more "clean," *thermonuclear* energy for electric power generating plants. Again, if mankind could find 800 billion dollars annually for weapons of destruction in the nineteen eighties, surely scientists and funding to finalize development of thermonuclear energy for everyday use could be found now. It is the "cleanest" source of energy, of which we will never run out of. Again, hydrogen is for all practical purposes an unlimited source of nuclear fusion from the oceans of our planet.

Since 1989 "cold fusion" has been a controversial subject. Up until now, fusion reaction required tremendous amounts of heat, such as it is created within all stars. However, some scientists swear by the theory of "cold fusion" reaction, others think it is nothing but a big hoax. It will probably take a few more years before mankind will know for sure, but the relentless research work by those who believe in it should be commended.

After all, when a few scientists work on an unconventional theory for over five years and they are not ready to give up, there must be something to it. Unfortunately, the idea is really so unconventional and contradictory, compared to known nuclear reaction that most scientists are not ready to accept it without more definite proof. What makes it even more difficult to accept for non-believers, is the fact that cold fusion does not create radioactive waste, the whole thing appears simply to good to be true.

Scientist Dr. Mills is working on the theory that hydrogen comes in two versions, one of them as yet unproven and with a much "tighter" orbit of the electrons around the nucleus. Maybe he is on to something?

Another controversial development is the theory of "nano technology" (one nano [millimeter] = one billionth [of a millimeter]). Scientists are already experimenting with this concept, with moderate success. Basically, it involves the arrangement of atoms in a desired shape, such as we can observe from plant growth for example. If the idea materializes, all manufacturing plants ultimately become obsolete, since every product right up to sophisticated machinery and automobiles can be "grown" in tanks.

It was said before, maybe in other words: Science in the twentieth century is like a new-born baby, discovering its surroundings. But we can rely on it, since it is based on fact and logic — more than we can say about ancient scriptures.

THE FINAL CHOICE IS OURS

Summing it all up, instead of concentrating on some of the priorities closest at hand, humanity is still involved in costly religious and territorial conflicts. In places such as the former Yugoslavia, Sudan and Afghanistan — lives are still lost over "traditional rights", fighting over territory much like our supposedly less intelligent "fellow" animals. By following old, unproven religious myth — Israeli settlers in the occupied areas claim their "biblical" right. Others are still enforcing senseless ancient traditions and are imposing ethnic "cleansing" and religious "superiorities."

Throughout the past chapters, I have tried to emphasize the importance of distinguishing between scientific "facts," "fact-based theories" and "twentieth century common sense," compared to usually inherited religious "faith" or "believing" in the priorities of senseless tradition. With what could be identified as "did you know that . . ." episodes, I have introduced a number of often misunderstood or ignored scientific discoveries and believable events.

Without having to go through about a dozen science text books, science publications, the Bible, the Koran, a stack of books on UFOs and having to watch a number of TV documentaries, the reader had the opportunity to draw conclusions between *"believing"* and *"accepting fact and logic"*. There may be a lot of "new" information to be found in these chapters for some, or not all that much for others. If only a FEW readers have a clearer view of scientific realities and theories on one hand now, and of major religions on the other, the author would consider the "mission accomplished."

Astronomy, biology, medical science, physics and chemistry may be a long way from being perfect and much more may have to be explored, discovered or even contradicted, but it is ALL WE HAVE to work with today and in the near future. To understand the world and the universe, to prevent diseases and to prepare for a better future, only science provides the answers sooner or later.

The "status quo" of religious world domination is no longer working nor acceptable, the educational and political systems of the world are out of touch with reality, predominantly influenced or dominated by religious, illogical and often harmful dictatorship.

Religion in most developed countries is on the decline, but billions of "believers" of one interpretation or the other of ancient scriptures still waste their lives, by letting themselves be domineered by ancient BELIEF, and totally ignoring the realities of the twentieth century . . . My goal was to eliminate as many doubts as possible, maybe to categorize our lives into "facts," "logic" or "belief," based on the title: "Reality before Belief."

If only religious leaders and politicians were to think in those terms, most conflicts could be avoided. Lives could be spared. Scientists would have funds available for research on important life-saving missions, rather than on life-taking technology. Why "believe" in something and act irresponsible because of it, when there is a better choice: known FACTS and LOGICAL theories, and above all: "COMMON SENSE?"

Understanding FACTS and theories of astronomy and all sciences MUST BE humanity's priority over ancient religious doctrine at the dawn of the twenty-first century. Why waste time in schools and churches on 3000 to 4000-year old illogical myth, when the reality is at hand or is still to be discovered?

I have made MY choice after looking at the facts and theories of the sciences and comparing them with the results of religious faith. As an agnostic, I find "spiritual" fulfillment and general satisfaction in scientific discoveries and in those theories, which are based on logic. And in the fact that I am one of millions who are more of a "beneficiary" of circumstances, than a "victim" of them. The preceding pages give the readers examples of what science has done and can do for us, and what major religions have not only NOT given to mankind, but also what they have taken.

When entire nations still live by illogical, outdated ancient holy books and base their laws on unfounded belief, when their governments ignore the realities of the twentieth century — the results will be the same as they were hundreds of years ago: Collective hatred towards all other "believers" and non-believers, sooner or later resulting in violence, death and destruction, with continuing acceptance of misinformation and non-acceptance of scientific facts. There must be millions out there, who think in the same terms, either before or after reading this book, and who want to accept scientific facts and theories BEFORE blindly accepting ancient scriptures, faiths and traditions as "fact." I would like to hear from them.

Ironically, just upon completion of the manuscript, one of the most significant archaeological discoveries was officially announced through the news media. After nearly a year of secrecy and extensive testing to confirm and re-confirm the calculations, the discovery of skeletal remains of the oldest common ancestor of homo sapiens and chimpanzee in Ethiopia was announced. The 4.5 million year old remains are indeed considered to be the "missing link," a common ancestor for man and ape. The facts of evolution have once more been significantly confirmed — a major defeat for steadfast religious anti-evolutionists, a triumph for REALITY OVER BELIEF.

As projected in the chapter on the Hubble Telescope, the strongest evidence ever about the existence of planets outside the solar system has now been confirmed. As an update on an earlier chapter, five developing stars in the Orion Nebula, about 1,500 light-years distant, have been found to be surrounded by discs of dust and gas, the material necessary for the formation of planets. Five billion years from now, when the solar system has expired, earth-like planets may be orbiting around any of THESE stars at the same stage of evolutionary development as Earth is today.

Along the same lines, Dr. Alexander Wolszcan discovered what appears to be three planet-sized objects in orbit around a "nearby" pulsar. His comment was, "If planets can form around strange objects such as pulsars, most regular stars must surely have planetary companions." The latest discoveries also verified the presence of water, the basic ingredient of life throughout the entire galaxy.

INDEX

A

C

E

F

G

H

I

J

K

L

M

N

S

ILLUSTRATIONS